TUMBLEWEED

TUMBLEWEED

Six Months Living on Mir

Shannon Lucid

Copyright © 2020 Shannon Lucid
All rights reserved.
MkEk Publishing
ISBN: 978-0-578-67109-3

DEDICATION

Kate, Amberle, Paige, Briana, Brooke, Caleb, Daniel, David, Skade, Talus

TABLE OF CONTENTS

PREFACE	1
THE BEGINNING	5
START CITY LIVING	11
TRAINING	41
LAUNCH	77
COMBINED OPS AND GOODBYE	87
"COSMIC" LANGUAGE	90
COMM AND ALL THAT	96
FOOD AND MEALTIMES	112

FORM24-OTHERWISE KNOWN AS COMMANDS FOR THE DAY	117
EARTH OBSERVATIONS	125
EVENINGS	130
PROGRESSES	137
PRIRODA	147
EXTRAVEHICULAR ACTIVITY (i.e, SPACEWALKING)	161
FORTY-FOUR BAGS OF S OUP TO GO	172
CREW EXCHANGE	176
STS-79, MIR DOCKING AND LANDING	188

PREFACE

It was my last Sunday in Moscow. In two days, all the Russian training for my extended stay on Mir would be complete. I would then return home, to Houston, Texas, to join the STS-76 crew in the final training for our launch that would take me to Mir at the end of the month. As I was leaving the movie theater, which, on Sundays, doubled as an international church, I bumped into a woman whom I had often seen during the year of Sundays that I had been attending this church but had never talked to before. After our exchange of "excuse me," she looked at me and then asked, "Are you the one I just saw on CNN? Are you the American woman who is going to spend four months locked up in a tin can with two unknown Russian men?"

Never having thought of my impending adventure in quite those terms, it took me a microsecond to realize that she was referring to the shuttle launch in three weeks that would bring me to Mir to join Yuri Onufriyenko and Yuri Usachov, who were already living and working there. I answered "yes", then headed out into the cold, gray, snowy morning.

On the metro and then the electric train headed back to Star City, where I was living and training for the upcoming flight, I had plenty of time to reflect on this woman's unique perspective. She was right in one respect: Yuri Onufriyenko and Yuri Usachov were, in conventional terms, almost unknown entities to me at that point. True, I had been in Russia for a year preparing for the flight to Mir, but most

Lucid

of the training had been classroom lectures on Soyuz and Mir systems. I had done a few simulations in the Soyuz and Mir mock-ups with Yuri and Yuri, but we had not had any time to get to know each other, to be able to interact with each other in unscripted, "normal" situations. Despite this, I was eagerly anticipating joining them on Mir as much as I would anticipate a reunion with family members.

There were two reasons why I was looking forward to living and working with Yuri and Yuri. The first was the result of a comment by Ellen Baker, a fellow astronaut and friend. After I told her who my crewmates on Mir would be, she spontaneously exclaimed, "Oh, Shannon! You're so very fortunate. You couldn't ask for greater guys to fly with." Over the years, Ellen and I had worked together, trained together, and flown together. We had learned to trust each other's opinions. She'd had the opportunity to get to know Yuri and Yuri while they were stationed at the Johnson Space Center (JSC), training to be the backup crew for the Russians who flew with her on STS-71. Her glowing one-sentence recommendation was all I needed to alleviate any apprehensions about living and working with Yuri and Yuri for an extended stay.

The second reason was the short conversation I'd had with Yuri Usachov in the Mir mock-up during one of our few training sessions together in Star City before the launch. We were working at the table in the Base Block, assembling a Russian medical experiment. In the middle of going over the checklist together, he looked at me with a grin and said, "You know, Shannon, if we were doing this on Mir, we would stop right now, have lunch, and look out the window at our planet Earth. When I previously flew on Mir, meals were the very best times. While drinking our bags of tea, we would gaze down at Earth and discuss life." I smiled back at him in agreement. At that moment, I knew we shared a common philosophy of space travel: that the most important thing about a spaceflight is not *what* you get to do but *whom* you get to do it with. It was then that I knew I was not worried about being "locked up in a tin can with two unknown Russian men."

During my six months on Mir, extended from the originally planned four months, I never had cause to rethink or amend my initial impressions. Reality proved to be a far richer experience than the anticipation had predicted. The general public often assumes that great scientific discoveries, rendezvous conducted with the utmost precision, and spectacular spacewalks are the memories carried home by most space voyagers. Certainly, astronauts do return with such recollections. However, the ones they revisit most often in the years following a spaceflight are those involving their crewmates. It's the little things that occur on a day-to-day basis, like the laughing exchanges over a bag of tea and the events that, in the retelling, end with, "I guess you had to have been there to understand how funny that was." Those are the moments that make the difference between a mundane trip and a great one.

The Mir 21 crew—Yuri, Yuri, and I—docked and reconfigured the final Mir module, Priroda. Together, we unloaded and reconfigured two Progress vehicles; accomplished five spacewalks, with my crewmates doing the actual spacewalks while I supported them from inside Mir; carried out the entire National Aeronautics and Space Administration (NASA) science program and even began the experiments scheduled for the next Mir flight; weathered extensive flight extensions; welcomed a new Russian crew, which included a French research scientist; and accomplished a crew exchange.

I often catch myself thinking of these exciting times, but many of the memories that I cherish most are of Yuri Usachov, such as him always saying "thank you for the great meal" (and meaning it!) as he floated away from the table at meal's end. Every morning, he would greet me with laughter and song as I drifted into the Base Block, ready for a new day. The song was always the first line of "Are You Sleeping, Brother John?"—the only English phrase that he knew. His young daughter had taught it to him as she was studying her grade-school English lessons. I remember his delight when he found, and then gave to me, a packet of my favorite Russian cookies. It had floated behind one of the panels of Mir, and we had long since given up hope of ever having any more cookies for dessert. Every day, there were also jokes about my interspersing English phrases like "you know" and "I mean" into my broken Russian. I would invariably reply that it was my duty to repeat standard English phrases so that my crewmates would have the opportunity to learn them, complete with an Oklahoma accent.

When I reflect on my six months on Mir, I have no shortage of memories. However, one event captures the legacy of the Shuttle-Mir program. One evening, Yuri, Yuri, and I were floating around the table after supper. We were drinking tea, eating cookies, and talking. The cosmonauts were very curious about my childhood in Texas and Oklahoma. Yuri Onufriyenko described the Ukrainian village where he

Lucid

grew up, and Yuri Usachov talked about his own Russian boyhood village. After a while, we realized we had all grown up with the same fear: a nuclear war between our countries.

I had spent my grade-school years living in terror of the Soviet Union. We practiced bomb drills in our classes, all of us crouching under our desks, never questioning why. Similarly, Yuri and Yuri had grown up with the knowledge that American bombers or missiles might zero in on their villages. After further discussion about our respective childhoods, we marveled at what an unlikely scenario had unfolded. Here we were, from countries that were sworn enemies just a few years previously, living together on a space station in harmony and peace. And incidentally, having a great time. Not long before, such a plotline would have been deemed too implausible for even a science fiction novel!

THE BEGINNING

In 1996, I spent six months aboard the Russian space station Mir. Though I wouldn't describe the experience as "better" than my four prior space shuttle flights, the long duration of the Mir mission provided me with more time to reflect on the actual experience. Almost every day I floated to the large observation In 1996, I spent six months aboard the Russian space station Mir. Though I wouldn't describe the experience as "better" than my four window in Kvant 2, a module in Mir, and gazed down at Earth below or into the depths of the universe. Invariably, I was struck by the majesty of the unfolding scene. But to be honest, the most amazing thing of all was that here I was, a child of the pre-Sputnik, Cold War 1950s, living on a Russian space station. During my early childhood in the Texas Panhandle, I had spent a significant amount of time chasing windblown tumbleweeds across the prairie. Now I was in a vehicle that resembled a cosmic tumbleweed, working and socializing with a Russian air force officer and a Russian engineer. At these times, I would think back on and contemplate the myriad of seemingly unrelated life events, both personal and in society, that had coalesced to create this experience.

I was born in 1943 in Shanghai, China, where my parents were missionaries. The summer before starting kindergarten, to escape the sultry heat of Shanghai, my parents decided to live in the mountain village of Kuling. My father traveled ahead by train, and a week later the rest of the family followed. We flew the first part of the trip in a

Lucid

World War II DC-3 that had been pressed into civilian use. The cabin was not pressurized. I distinctly remember my mother turning green and my brother and sister throwing up as we rocked and bumped over the ridgelines of the mountain ranges. I sat transfixed, nose pressed to the window, staring at the scene beneath. As the plane started to descend, I saw a tiny strip of gravel. At one end stood the very small figure of a Chinese man with a red scarf wrapped around his neck. It was beyond my comprehension that the pilot at the controls of the airplane, a mere mortal after all, could land on that small strip. It was at that moment that I decided I would be a pilot when I grew up. The next airplane I was in was an Aeronca Chief. I was nineteen, taking my first flying lesson.

During that Kuling summer, I was fascinated to learn that when it was day in our mountain village, it was night back in my grandmother's Texas. I puzzled over this fact, then decided to find out if it was true. I asked my frazzled mother, dealing with some crisis I no longer remember, if I could walk around the world to see what happened to the sun on the other side. Her distracted yes was all I needed to start off. By nightfall, I was surprised to find I was still in China. Extremely tired, I curled up on a warm rock to sleep and wait for daylight. The next day, I would finish my quest for the secret of the sun's movement. My frantic parents found me fast asleep, but no amount of logic could budge their firm no to my incessant pleas to continue my journey. "Just wait," I told them. "When I grow up, I am going to explore the entire world!"

An avid reader, I devoured every book I could find. In the fourth grade I stumbled on a book that told me that water was composed of hydrogen and oxygen—two gases. How could this be? How could two gases make a liquid? This fact seemed marvelously beyond comprehension. Finding out that it was chemists who discovered and understood such miracles, I decided that not only would I be a pilot and an explorer, but I would also be a chemist. The more I read, the more I worried. I was so young. The entire planet would be explored before I could get out of grade school. By the time I became an adult there would be nothing left for a chemist or an explorer to discover. Then I stumbled across a book describing Robert Goddard's work with rockets in the New Mexico desert. My dreams took a quantum leap outward. I would not just be an explorer of Earth but a scientist

Tumbleweed

who explored the universe! This would allow me to combine all my desires: flying, exploring, and making scientific discoveries.

Now there was no limit to the dreams a girl could have. I read books. I built rockets and models. Using a cardboard box left over from the installation of a new hot water heater, I made a "rocket" to go to Mars. I put it in the attic, then spent a day sitting in it, training myself for the long journey across a great expanse of nothingness toward Mars. At the end of the day, with my meager food supplies depleted, disappointed that I had not been missed, I headed downstairs toward the kitchen and family, deciding to delay training until another day.

In the eighth grade, when assigned the task of writing about a career, I chose "rocket scientist." Not only was this something I really wanted to become, but it had the added advantage of not having any information on it in the library, so I was not hindered by facts when writing my paper. I could just create the requirements, the working conditions, and anything else I wanted for the assignment. My teacher, markedly underwhelmed by my choice of careers, admonished me to take her assignments seriously.

Growing up, it seemed that no one else shared my dreams of exploring the universe. I could not understand why the United States was not already building colonies in space. Then, I read a short article in the newspaper about the Russians planning to launch a person into space. I cut out the article and showed it to my parents, saying, "I guess a person will have to become a communist if they want to live on Mars!" My parents did not respond well to this un-American, unpatriotic statement. I must admit that I often contemplated the eerie foreshadowing of this incident while on Mir.

Sputnik was launched during my sophomore year in high school. Space exploration by Americans, though conducted more to outdo the Russians than to explore anything, was finally a reality. I finished high school, went to college, majored in chemistry, and learned to fly airplanes. I read everything I could find about space exploration, only to realize that it did not matter how much I learned or what I could do—there was no place for a female in the space business. The seven original all-male American astronauts grinning on the cover of *Time* magazine were the final proof. With the desperation of total exclusion, I wrote to *Time* and asked why only males were part of the American dream to explore space.

Lucid

Social changes were sweeping across the country faster than the American space program was sweeping through the universe. After American men walked on the moon, after American men inhabited Skylab for a record eighty-four days in space, after the space shuttle was designed, these social changes finally collided with the space program. By governmental decree, a person could no longer be excluded from a job just because they were female. This meant, by law, that females had to be included at all levels of government jobs, including those of the space program, even including the job of astronaut. I read a few sentences in *Science* magazine stating that NASA was considering including women in the next astronaut selection process. I responded to the article by immediately submitting my application to NASA. Even today, I still find it hard to believe that I was part of the astronaut class that included the first females.

After 1974, when the Skylab program ended, America turned away from long-duration missions to focus on the short-duration missions of the space shuttle. The Russian space program, however, continued to focus on expanding the length of time cosmonauts stayed in low Earth orbit, first on the Salyut space station and then on the station Mir. By the 1990s, as the Cold War between America and Russia was rapidly becoming a thing of the past, it seemed only natural that the US and Russian space programs should cooperate in the next big step of space exploration: the building of the International Space Station.

The first phase of this cooperation was the Shuttle-Mir program. It was agreed that one astronaut would fly with two Russians aboard the Soyuz, the Russian vehicle that transports crew members to and from Mir. A shuttle flight would then dock with Mir, and the crew, one American and two Russians, would return to Earth on the shuttle. Norm Thagard, a medical doctor, engineer, former naval aviator, and veteran of four space shuttle missions, was selected as the first NASA astronaut to serve aboard Mir. Norm was perfect for the assignment.

Unlike space shuttle crews, all Russian crews had a backup team that trained concurrently with the primary crew. Bonnie Dunbar, a veteran of three shuttle flights, was selected to be the backup for Norm.

After Norm and Bonnie began training in Russia, the Shuttle-Mir program was extended. NASA added a series of shuttle flights to send American astronauts to the Russian space station. Each astronaut would stay on Mir for approximately four months, performing a wide range of peer-reviewed NASA science experiments. The space shuttle would periodically dock with Mir to exchange the American crew member and deliver supplies. In addition to the science, NASA's goals were to learn how to work with the Russians, gain experience in long-duration spaceflight, and reduce the risks involved in building the space station.

My direct involvement in the Shuttle-Mir program began in 1994. At that point I had been an astronaut for sixteen years and had flown four shuttle missions. Robert "Hoot" Gibson, then the chief of the astronaut office, called me late one Friday afternoon. He did not waste time in small talk. After the initial hello, he said, "Shannon, are you still interested in flying on Mir?"

"You bet! When can I go?"

"We haven't gotten approval to assign a crew to the upcoming Mir missions, but time is getting short. We can't wait forever for potential crew members to begin studying Russian. We have approval for you and John Blaha to start studying Russian at the Defense Language Institute in Monterey, California. Now, I need to caution you. This does not necessarily mean that you are going to Russia. This is not a crew assignment; it is only an assignment to study Russian. Our hope, our plan is that you will go to Russia to train and eventually fly on Mir, but none of that has been approved. Can you fly out to Monterey to begin Russian lessons on Monday?"

"Well, Monday would be a slight problem. My daughter is getting married next Saturday, and next week I'll be busy getting ready for the wedding. How about if I go out a week from Monday?"

"I guess that's OK. We've waited this long. What difference will one more week make?"

"Thanks so much. I am just so excited about getting to do this!"

"Well, congratulations. I hope this works out so that you are assigned to a crew and get to fly on Mir!"

I hung up the phone. For a few minutes I just sat there, basking in the warm glow of the anticipation of another spaceflight, and a long one at that. Then, reality started to set in, and I began to wonder just what I had gotten myself into. I called my husband, Mike, who was at work, to let him know. Over the phone I could imagine his eyes rolling heavenward as he wondered what new adventure I had gotten us into. Then I went to find John Blaha to compare notes about our conversations with Hoot. I had flown two shuttle missions with John

Lucid

as the commander and had always enjoyed working with him. It would be good to jump into this adventure with a friend. After I found him, he told me he'd received basically the same message from Hoot: no assignment to a crew, no assurance of going to Russia, just an opportunity to learn Russian.

At this point it would be fair to ask "why". Why did I want to learn a new language? Why did I want to leave home for almost two years? After all, I was already on the mature side of middle age. Then again, this seemed like a perfect time. Our daughters, Kawai and Shani, had already graduated from college and were self-supporting. Michael, our son, the youngest, was away at college. Our homelife was almost stress-free. This was an opportunity for a long-duration spaceflight. This was too great an opportunity to pass up. I was really, really curious to discover what it would be like to live in space as opposed to just a short visit like on my shuttle flights. For a scientist who loves flying, what could be more exhilarating than living and working in a laboratory as it hurtles around the Earth at seventeen thousand miles an hour?

John and I arrived at the Defense Language Institute in August of 1994 to begin studying Russian. During our stay in Monterey, neither John nor I ever heard one word about whether we would be going to Russia. Optimistically, John went ahead and bought a coat suitable for the Russian winter. I held off, not wanting to invest that amount of money in clothing unless I was certain to use it. Toward the end of October, I received a phone call from someone in the travel office back at the Johnson Space Center wanting to ask me a few questions. I asked her why she needed to know that information. She replied— the surprise that I needed to ask evident in her voice—that she was processing the visa for my trip to Russia in January. With that piece of information, I went out and bought a down-filled jacket and boots good for walking on ice and through the snow. I was ready for my Russian adventure to start!

STAR CITY LIVING

Peering through a small portion of the grimy bus window, smudged semiclean with my gloved fist, I caught the first glimpses of my new home. Star City—the name said it all. A place of myth and legend. It was where the cosmonauts prepared for space travel, where Yuri Gagarin had trained for his historic flight, humanity's first tentative step into the cosmos. What I saw through my small peephole reinforced this aura of mystery because spread out all around was an enchanted winter wonderland. Burdened by blankets of white snow, tall black pine branches sloped downward, almost touching the huge banks of snow sparkling in the moonlight.

Home Sweet Home

As the bus pulled into a circular drive, John and Brenda Blaha and I stood in front of a ten-story rectangular apartment building called Dom Cheteria, or in English, Home Four. In whatever language, it was to be my home away from home for the next year. (Brenda had moved with John to Star City for the year, but Mike had to remain in Houston because of work. He used his six weeks of vacation to come to Star City at various times.)

We slithered across the icy steps into a dark entryway. There, we squeezed into a four-person elevator, rode up to the fourth floor, and entered my apartment. The large entryway, which proved essential for

Lucid

shedding wet winter coats and boots, led into a combined living/dining room, newly furnished in a vaguely Western American style. I was surprised to discover two bathrooms, although each could only be accurately described as a half bath. The tub was in one, and the toilet was in the other. (Both contained a sink.) The larger of the two bedrooms was totally filled by a double bed and wardrobe. The smaller second bedroom contained a small bed and desk. I immediately designated this smaller bedroom as my study.

The ceilings in the apartment were high, and the floors were genuine wood. The living/dining room and larger bedroom had huge windows. Glass doors in each room opened out onto a small balcony. This balcony area was essential in Russian apartments, which generally did not contain clothes dryers and had no access to a usable outdoor yard. Without a balcony, there would be no place to dry clothes.

My first impression (which proved accurate) was that this would be a cozy, comfortable place to live. I soon discovered, however, that my "cozy" apartment was considered luxurious by Russian standards. Every Russian visiting my apartment for the first time would look around in wide-eyed amazement and invariably ask, "You live here all by yourself? All this space is just for you?" Wonderful questions considering that the typical Russian apartment was just one room in which three generations plus a large dog (and maybe a cat) lived.

The apartment did have its idiosyncrasies. All apartments in the building were heated by circulating hot water delivered by pipes from a central heating plant. The temperature was controlled from this central plant and could not be changed by the inhabitant. The Blahas' apartment directly above mine was always very hot in the wintertime— so hot that sometimes they had to open the door to their small balcony to regulate the temperature. My apartment, however, was frigid. Every night while studying, I would sit bundled up in heavy bathrobes and blankets with only my nose exposed. One night while intently reviewing the vocabulary for a test the next day, I heard several popping noises emanating from the kitchen. Surely I had not heard gunshots! Trailing all my blankets behind me, I rushed to investigate. My "gunshots" proved to be three cans of Diet Coke that had exploded upon freezing. (Diet Coke was a precious commodity. It could be found only in the American Embassy commissary.) Not having time to clean up a mess that night, I just put the cans plus their frozen contents in the sink. The next morning, the Coke was still frozen. The kitchen had never warmed up to the point where they thawed.

Like the heat, hot water availability was also regulated by the calendar. Every year, all the hot water was turned off for the month of August so that the pipes at the heating plant could be cleaned. Kawai, my oldest daughter, came for her first visit during the second week in August. Every morning, as she shivered through a quick ice-cold shower, she reminded me of my failings as a mother. A *good* mother would have suggested that her child visit in July, not August.

Star City

Star City is a self-contained town located in the middle of a forest northeast of Moscow. A high concrete-block wall encloses the entire area. Russian soldiers guard the front gate. (I never saw any soldiers at the back or side gates.) All vehicle traffic going into the city must be waved in by these soldiers. In the early days, there were many instances of NASA people sitting in their transport van for hours, trying to get the paperwork straight so they could get into the city. I always walked in or out on foot or on my bicycle and never once was asked for any type of identification.

Inside the perimeter wall is another wall with a gate dividing Star City into two parts: the living area and the working area. At the time of my stay, the living area consisted of high-rise apartment buildings; the local school; the Cosmonaut Center, which was a large auditorium and a space museum; a small lake; and a few shops where bread and a sporadic selection of groceries could be bought.

Located in the center of the apartment buildings was a big grassy parade area with a walkway around the perimeter. During the summer, when my son was visiting, he spent the day watching, with great fascination, a couple of young soldiers mow this vast expanse with a push manual reel lawn mower. That evening, he asked me what I thought they could have done wrong to merit such punishment—using a manual push mower instead of a riding mower. No punishment, I said—that was just the only equipment they had to get the job done. "Oh," was his unbelieving answer.

NASA had a small office in this part of Star City. The person who ran it was called the DOR, Director of Operations, an astronaut assigned to carry out administrative duties and to be the NASA interface with the Star City Russian bureaucracy. This assignment was

Lucid

a six-month rotation. The NASA office was located in the Prophylactory (a.k.a. the "Prophy"), a three-floor dormitory built by the Russians to house Americans during the Apollo-Soyuz Test Project. When NASA returned to Star City for the Shuttle-Mir program, it leased the second floor as an office and living space for the DOR, the supporting flight surgeon, and three professional young Russian women without whose work as office managers, translators, and cultural brokers NASA wouldn't have been able to accomplish anything.

These folks worked hard and long hours as they tried to maintain some type of contact with their counterparts back at the Johnson Space Center while coping with an eight-hour time difference. John and I had very little contact with them. We were always in some type of training or in our apartments studying. Sometimes I could go for several weeks without speaking English to anyone except John as we walked together back and forth to class.

On the other side of the wall was the working area. This was where all the training took place. In the morning, just before nine, and after lunch in the afternoon, just before two, John and I, along with all the working people of Star City, streamed into this section. Twice every day, this small river of people also left the area, at one and at six. The gates were always guarded by a young military person. Not once was I ever asked to present my *propusk*, my identification. I assumed that we were known by sight.

One day, though, we left after six o'clock due to some last-minute medical tests. As we left the building, we scurried to the gate. We worriedly mumbled through mouths stiff with cold about what we would do if the gate were already locked. We arrived at the gate. It was locked tight. The guard was gone.

We exchanged glances, wondering what to do now. Our mouths were too frozen to discuss our situation, so we simply shrugged in resignation. Just then, a Russian woman came up behind us. Seeing our predicament, she motioned us to follow her. We left the shoveled path and followed her down a well-beaten trail stomped in the icy snow. We came up to the fence and watched in amazement as she got down on her hands and knees and went under a gap in the wall that was hidden from view. With another shrug of our shoulders, we followed suit. Once on the other side, we just looked at each other and laughed. The next time we were on the wrong side of the wall after six, we did not give it a second thought. We were locals now. We knew about the fence gap!

The security of the military airport, Chkalovsky, located adjacent to Star City, had the same type of "Potemkin village" aura. Entering on a Star City bus through the well-guarded front gate always took some time. The papers of every person were scrutinized by the guards. The reverse process, trying to depart from the airport to return to Star City, was no less time consuming. It could take several hours after landing to exit the airport. Our papers would be taken inside some building to be scrutinized again while we sat stranded on the unheated bus in the gathering darkness, shivering in the winter cold.

The strange part of this system wasn't the security measures themselves. Certainly, tracking who enters and leaves a military airport makes a lot of sense. The odd part was that all this security was just if you wanted to use the front gate.

During one of my son's visits, we were riding our bikes on paths through the surrounding forest. The trees ended abruptly, and Michael almost fell off his bike onto the visual approach system at the end of Chkalovsky's main runway. I took his picture as he stood there, touching the runway light and hanging on to his bike. Then, we lay on the path at the end of the runway and took pictures of the bellies of the huge military transport planes as they flew over our prone bodies just before touching down.

Calling Home

That first night, the few Americans already living in Star City welcomed us to our new apartments with gifts of instant oatmeal and chicken soup. As soon as they left, my first thought was to try to call home so that Mike would know I had arrived safe and sound in Star City. I was also eager to describe to him my new home for the next year. I got out my calling card and dialed straight through to the United States. Mike answered, and his voice was crisp and clear. Twenty minutes later, I hung up with satisfaction over the clear connection, little suspecting this was the last time calling home would prove to be so simple!

During my year of living in Star City, keeping in touch with my family back in Houston was at the top of my personal priority list. It was also the source of endless frustration. On subsequent evenings, I tried to call home. Sometimes it would take up to forty-five minutes to get an international operator. Then, the connection would go dead. If

Lucid

I were finally patched through, the voice on the other end would be so faint or drowned in static that we sometimes simply hung up on each other totally exasperated because we could not hear.

Fortunately (or so I thought), I had never intended the telephone to be my primary means of communication with my family. Back at the Johnson Space Center, we had been told that it would be possible to send and receive email at the NASA office in Star City. I had faithfully carried the procedure on how to get connected all the way over to Russia in my pocket. The good news is that I arrived with it in place and intact. The bad news is that, due to massive system inadequacies, I was unable to establish a connection and send or receive any email.

Every night, with ever-decreasing hope of success, I loaded my laptop computer into my backpack. Then, depending on whether the paths were covered with snow or with ice, I spent fifteen to forty-five minutes hiking from my apartment to the NASA office. Once I arrived, I tried to establish a connection so that email could be exchanged. On all but two or three occasions, I finally stopped after a couple of hours of utter frustration. Each night, I repacked my equipment and negotiated the icy paths back to my apartment, completely dejected because of mail not received.

As I returned to my apartment, it was not snowflakes that I saw falling around me but all the ones and zeros that were clogging the ether—binary digits that had been trying to get into my computer without success. As they fell silently to the ground all around, I caught them on my gloved hands and fantasized about arranging the flakes in the proper order to read the messages I had not been able to receive.

Norm and Bonnie had already been living in Star City and were busy preparing for Norm's launch. We got together when we could. On one such evening, the topic of conversation was our total frustration at not being able to communicate with our families back in Texas. Bonnie told us that she had a phone connected to a satellite dish in her apartment and had no problem connecting to Houston. She graciously offered to share it with us, so I changed my nightly routine. Each evening, I would hike over to her place and use the equipment located there.

As Norm and Bonnie's departure from Star City was rapidly approaching, John and I had some long discussions about whose apartment, his or mine, Bonnie's gear should be moved to. We were excited about having direct access to this great equipment and were more than willing to share. I did not think, though, that being dependent on this one method of communication to reach the outside world was adequate.

In one of my nightly email messages home, I explained my dilemma to my daughter Shani, who has a master's degree in computer science. I mentioned that someone at the church I attended in Moscow had told me that CompuServe worked great in Russia and that I should be able to connect over the local phone lines to get email. I urged her to figure out how to do this. The next week, a person traveling from Houston to Star City brought me a package with CompuServe software and very detailed "mom-proof" installation instructions.

Shani's instructions worked. I got connected. Watching that first batch of mail arrive on the computer in my apartment felt like a miracle. Finally, I had a reliable method of communicating with my family.

It turned out to be a good thing that I had planned ahead. Norm launched, and Bonnie immediately went back to Houston to train for the shuttle flight that would bring Norm back home from Mir. All the communication equipment disappeared from her apartment. John and I never knew what happened to it. All we knew was that it never appeared in either one of our apartments.

Thanks to my personal CompuServe account, I didn't really care. Every night and every morning, I was able to connect. Admittedly, this connection process could take a long time—I sometimes had to try for over forty-five minutes to get a connection that would last for the three or four minutes it took to collect my mail. But in the comfort of my own apartment, this did not bother me much at all, especially compared with the long hikes I used to have to make to the NASA office. Now, I could sit on the couch wrapped in my blankets while I studied and just reach over and hit the appropriate key to get the computer to try, try, and try again until a connection was made.

No News Is Good News?

It was always difficult for the folks back in Houston to comprehend just how isolated we were, living in Star City, from the world and world events. True, I did have a Russian television, but it received only the Russian stations, and the news on these stations was just "talking heads." Trust me—Russian talking heads are like none you have ever

Lucid

seen. The mouths of these journalists travel at the speed of light, disgorging incomprehensible Russian. To make matters worse, there was very little video footage of the news events to help the viewer deduce what the subject was. When I despairingly told my Russian teacher that I thought I would never understand them, she, to my dismay, agreed with me. Then she said not to worry—much of the Russian populace could not understand them either!

Such programs were useless as an information source. I tried using my shortwave radio but was never able to find a reliable English news source. Additionally, my CompuServe connection would never stay up long enough to be able to surf the internet to find news. NASA tried to keep us informed by subscribing to the Sunday *Houston Chronicle* for us, and it was supposedly mailed to us in care of the NASA office at the American Embassy in Moscow. It was a rare paper that ever made it out to Star City. And of course, the ones that did were weeks old. Our guess was that someone in the embassy got to the NASA mail first and swiped them, but this was just a theory. There was never any proof. Why they never showed up—at least on a semiroutine basis—remained a mystery.

We did not need newspapers or television to know that the Russian-Chechen conflict was going on—instead, we had a shaking apartment building. At the height of the conflict, we could determine how the Russians were faring by how violently we were rattled in our beds as transport planes bound for their nightly Chechen raids took off from Chkalovsky. The planes skimmed over the top of our apartment building, which then shook in direct proportion to how heavily loaded they were.

We also had visual evidence of the conflict's progress every time we traveled down the road just outside the Star City wall. Located on the side of the road, across from the entrance to Chkalovsky, were a few scraggly shrubs. They became a "people's memorial." Strips of orange and black cloth fluttered forlornly from the bare branches. Around the base of these bushes were framed photos of uniformed young men. The strips were in memory of local youth killed in Chechnya. The number of photos and colored strips grew exponentially over the weeks. This was a more poignant testimony to the misery being endured in the conflict than any statistics-filled newspaper article.

The perfect example of how isolated we were from worldly events, however, was a cryptic email message I received early one morning

Tumbleweed

from my husband. The brief message simply said, "Don't worry. Joe is alive and well."

"OK," I thought. "Now just what could Mike mean by this? He certainly isn't in the habit of sending me daily health reports on my brother!"

I sent a note back. "Why did you think that I was worrying about Joe?"

(Remember, it took a day for each of these exchanges.)

"Because he works downtown in Oklahoma City, in the Federal Building!"

"So? He has worked there for years!"

"Haven't you heard? The Federal Building was blown up this week, and hundreds of Oklahomans are dead! I just wanted you to know that Joe had decided to go to work later that morning and wasn't in the building when it was bombed."

As I said, we were very isolated. My family finally understood my need for news and started to dribble bits of world events to me in their daily notes.

Mail

The Russian mail system was unreliable, so all our letters were sent to the NASA office located in the American Embassy in Moscow. The DOR was charged with making sure the Star City mail was picked up there once a week and brought out to us. It was always the high point of the week for me to go over to the NASA office and pick up my mail on Friday evening. Then I would take it back to my apartment and spend the evening reading the cards and letters so faithfully sent to me by my family and friends. That is, I did this on the good Fridays. Many times, I would leave the darkened, closed-for-the-weekend NASA office disappointed because the mail had not been picked up.

This all came to a head one Friday evening, just hours after my daughter Kawai had arrived on her first visit to Russia. Wanting to make sure that she stayed awake until dark to ease the time transition, I suggested an after-supper walk over to the NASA office to get my mail. Reading mail from home together would be a satisfying end to a long, frustrating week.

We got to the office, but there was no mail. Just then, the DOR walked into the office, and I asked if anybody had gone into Moscow that day. The answer was yes. As a matter of fact, the NASA van had made the trip—which took well over an hour—to the embassy on various errands four different times that day.

As you might imagine, the conversation slid downhill from there as I tried to fathom why not once, out of those four trips, the mail could have been picked up.

"Then where is the mail?" I asked.

"Oh, I guess no one thought to pick it up," replied the DOR. "But don't worry. We'll get it next week. Mail is no big deal."

With that, I turned and walked out. On the way back to my apartment, my daughter commented, "What you need over here is a social worker running the office and not an astronaut." A very astute observation.

On Monday, I skipped lunch to make a special trip to the NASA office. Upon arriving, I told Natasha, the Russian office manager, that I wanted to reserve the NASA van every Friday for one trip to the embassy.

Natasha started to write down my request, and then she looked at me and said, "But aren't you in class every Friday?"

I replied, "Oh yes, but I am entitled to use the van. I don't have to be in it. I'm just reserving the van every Friday to go down to the embassy to pick up the mail."

Natasha, super efficient as always, laughed and made sure the DOR had the mail picked up every Friday after that.

A Different World

Not only was it hard communicating with the outside world and the local NASA personnel, it was also difficult knowing how to initiate conversations with the people living in Star City. The first time I walked around Star City in daylight, I was struck by the fact that no one looked you in the eye. Because I arrived in the deepest part of winter and all the paths and roads were covered in snow and ice, my first thought was that everyone was busy watching their feet so they didn't slip. No one wanted a broken leg! Once the snow melted and still no one met my gaze, I thought, "OK, we have to keep our eyes downcast so as not to have mud-soaked shoes." Then, suddenly, it was overwhelming spring, with a brilliant blue sky, birds singing in the towering firs, and dry paths to walk on, and still, no one looked up. I finally concluded that it wasn't the ice or mud; not meeting a stranger's eyes was just a Russian characteristic. The difference must be cultural.

Tumbleweed

Maybe it was a form of self-preservation, the result of long years under communist rule, when your neighbor's business was definitely not your business. Whatever the reason, the lack of eye contact and a friendly smile was a constant reminder that I was no longer in the heart of Texas.

Another characteristic I quickly noticed was that in Star City, there were no newspapers or community bulletin boards to inform the citizens of upcoming local events. All information traveled by word of mouth. Surprisingly, this means of communication was extremely effective. I was always taken aback when someone I didn't know knew when my son had arrived, that it was taking a long time for my clothes to dry on the outside balcony, or that I had lima beans once again for supper.

The only community event that occurred while I was in Star City was a gathering to honor the cosmonauts and Norm, after they returned from Mir. The entire community came out marching, along with a military band, behind the newly returned heroes. The first stop was the statue of Yuri Gagarin that was the focal point, the northern star, of Star City. (When giving directions in Star City, you always began with, "You know where the statue of Gagarin is, don't you?") I noticed a group of young schoolchildren carrying huge bouquets of flowers. My initial thought was that giving flowers to Norm and his crew was a nice gesture. The children, though, did not even look at the live cosmonauts standing so close to them. Instead, they kept their eyes fixed on the statue of Gagarin, reverently laying their flowers at its base.

Nothing special, in the way of community events, ever took place in Star City to mark the weekend. There were no children's sporting events, no church services, no social events. There were no movie theaters and no cafés where you could meet your friends. The only difference that set Sunday evenings apart was that in the spring, summer, and fall, the citizens would dress up in their best clothes, the teenage girls in their nylons and high heels and the older couples in their suits, and promenade around the big, open grassy area located in the middle of the city. Everyone was dressed up, but there was no place to go.

Lucid

Groceries

Accomplishing the ordinary tasks of daily living took an inordinate amount of time in Star City. There was no one central point where most of my shopping could be done. Scrounging for the necessities of life and the common things I had always taken for granted—like bread, eggs, cheese, and fresh vegetables—was like participating in a permanent scavenger hunt. My son was the first one to arrive for the planned family rendezvous in Russia for Christmas. I asked what he would like for supper, and he told me. I replied, "I think I can do that," and we set off in the blowing snow to find the different parts of a normal meal—meat, bread, potatoes, and a few carrots. By this time, I had gotten fairly good at knowing the most likely places to find particular items. Two hours later, we had finally found everything for our simple supper. As we unlocked the door to my apartment and started to shed multiple layers of clothes, Michael said, "Wow! That was a lot of work for only a little supper for the two of us!" He had exactly comprehended the major frustration with living in Russia. It just took so long to do the common, everyday things.

Our lives were made a little easier in this regard because NASA had sent us over to Star City under the regulations governing "permanent change of station." That meant that we were allowed to ship a certain number of personal belongings when we moved. Before leaving for Russia, I had gone to a local discount grocery store and bought box after box of all the canned fruit and vegetables I thought I might eat during the course of a year. I also stocked up on toilet paper, paper towels, cleaning supplies, soap, shampoo—anything I thought I would not be able to live without and might not be able to find locally. These shipped supplies covered my basic needs, but a woman does not live on cans of creamed corn and lima beans alone, which is what I had two to three times a week. Bread is very nice. Fresh strawberries and fresh bananas are even nicer. And I won't even mention carbonated beverages!

Church

Before leaving for Russia, I spoke with a friend at the Johnson Space Center, Travis Brice, who had facilitated Norm's move to Star City. I asked him particularly about living conditions there and what to expect. My first question was, "On Sunday, do you think there will be anyplace for me to go to church in Star City?"

Tumbleweed

Travis looked at me and said, "Surely, you jest!"

I replied, "I know, but I don't think that I can live anywhere for a year without going to church on a regular basis."

I did not know how or where, but right then, I decided that my first goal upon arriving in Star City would be to find somewhere to worship regularly every week. The second week I was in Star City, I read a small article in the English Moscow newspaper that someone had picked up at the embassy and left in the NASA office; it described a group of international people who met every Sunday morning in a downtown movie theater for a church service. The next Sunday, I made it a priority to find the meeting place.

Upon entering the "church," I discovered that the congregation consisted of many different nationalities—one Sunday, there were over twenty-two countries represented—who shared a common belief and spoke English. Many were international students from African nations who were stranded in Moscow after glasnost. They had received scholarships to study in Moscow, but after glasnost, the funding disappeared. Unfortunately, they did not have the money to return to their countries of origin. As a result, they were scrambling to make a living in Moscow while trying to remain in school. There was also a group of Russians who came primarily because they wanted to improve their English skills, and there was a small contingent of Americans who worked at the US Embassy. My son, attending church with me during his visit, commented, "This is what I think a church should be like." I asked him what he meant. He replied, "Well, church should be a place with lots of different kinds of people, not just one color like ours back home, and it should also be a place where you can just come in your jeans and T-shirt!"

My Sundays rapidly fell into a pattern, one that I looked forward to all week. Every Sunday morning, I left my apartment precisely at eight o'clock. In the winter, I would bundle up against the cold in my down-filled winter jacket, long underwear under my corduroy pants, and heavy boots, which helped me keep from sliding on the ice. I hiked across the road to the well-trodden trail through the forest that led to the side gate of Star City. Generally, huge snowflakes would be falling out of the dimly gray, indistinct morning sky. At the entrance, I walked past the guard—looking down at the ground, of course—and joined the throng of local citizens headed down an asphalt path through more forest to the train platform. I carefully clambered up the crumbling

Lucid

steps and then, stepping over huge cracks in the decaying platform, stooped down at the window of the ticket agent and requested a ticket to Moscow and back. After that, I zipped open my backpack, carefully placed this ticket into the book I was currently reading, and waited for the train. There was no conversation among the waiting people. The silence was so profound I could almost hear snowflakes slither and slide through the frozen air. When winter finally started to recede, the sounds changed. Each week, standing there on the train platform was a new experience because I heard new species of birds lustily singing as spring rapidly advanced.

Upon boarding the train, I tried to sit in a car that contained family groupings and, most importantly, had no broken windows. Because Star City was near the beginning of the line going into Moscow, there was never a problem finding a seat. I would settle into a window seat and pull out my book from my backpack. This was my weekly free time. It was my gift to myself, a time to read without feeling guilty because I was not studying Russian. With my book in my lap, I fit in with the Russians on the train, most of who would be reading. Russians always seemed to be reading when waiting or riding on the train or the metro (i.e., the subway). Every time I saw this, I thought to myself, "These are my kind of folks!" Russia—a country where you could stand with an open book while waiting for the next happening in life and not feel out of place.

Every few minutes, the train would lurch to a stop, and more people would get on. By the time we reached the Moscow station, it was standing room only. At many of the stops, peddlers would get on and shout information about their wares, which might be books, newspapers, ice cream, or miscellaneous parts to a hand-operated meat grinder. Only rarely did a passenger purchase an item.

In addition to the peddlers hoping to make a few rubles, itinerant musicians would sometimes stroll down the relatively empty train aisle during the early part of the trip. One bitterly cold Sunday morning, a clarinet player waltzed by, playing a fast-paced polka with a round Cheetos can tied to his waist to catch tossed rubles. I pitched in a few, and he grinned back at me, never missing a note.

Looking out the window, I was always mesmerized by the passing landscape. The word *bleak* took on a whole new meaning for me. We rolled past the disintegrating infrastructure of a nation—factories that were now nothing but rusted hulks, deserted and decaying apartment

Tumbleweed

complexes, crumbling concrete of former roads and bridges. Looking around the train, I often thought, "How can these people find a means of existence in the midst of all this disintegration?" All markers of a human presence—buildings, factories, bridges—were relics of the past; nothing was present or future, unless you counted the vegetable gardens.

These microgarden plots were located along every track and were tenderly cultivated by individual Russians so that they might have a few vegetables to supplement their diets. The plots were fenced with junkyard treasures, rusted car doors, tin roofing removed from the rusted factories, and old planks of wood. Inside these folk-art fences were neatly tilled rows of vegetables, baby plants protected in early spring with scavenged plastic sheets. During the long summer evenings, Russian families would gather in these gardens, sitting and socializing around small smoking fires that helped keep the mosquitos away.

Arriving in Moscow, I walked half a block to the red-line metro station (the metro lines are differentiated by color) and then shuffled my way into the ever-flowing river of people slowly being funneled down to the escalators, which briskly moved the human flood down to the earth's bowels, where it would then be a one- or two-minute wait for a metro train.

I never ceased to marvel at the metro in Moscow. Sure, it was old, and many of the cars were decrepit, but entering a metro station was like visiting a museum. Each station had a different theme, illustrated by artwork on the walls and ceiling. Of course, there were the mandatory statues extolling war heroes, but there was also a station named after the Russian writer Anton Chekhov, dedicated to a literary theme, and a Mendeleev station, named after the creator of the periodic table, which displayed various molecular structures suspended from the ceiling. I never ceased to marvel at a people who applauded literary and scientific accomplishments in such a public forum.

Each Sunday, I transferred from the red to the brown line, counted five stops, and then got off. Next, I walked half a block to the movie theater and went up to the fifth-floor auditorium. I always entered the door at ten on the dot. It took exactly two hours to reach church from my apartment. At noon—the church had to end promptly at noon so that the space could be used for the afternoon matinee—I left the

Lucid

theater building. Then I would carry out whatever plan I had conceived the night before.

Sunday Afternoon Explorations

Every Saturday evening after I had studied Russian until I could do so no more, I called it quits and pored over my maps and Moscow guidebooks, planning an excursion for the next day. Sunday afternoon was dedicated to the exploration of Moscow.

It took me two Sundays to locate Leo Tolstoy's Moscow residence. On the second Sunday I spent searching for it, I was elated when I walked up to a gate and saw the Cyrillic sign that read "Tolstoy's Home" and then listed the hours it was open. "Great," I thought, "it's only three thirty, and it doesn't close until four thirty. I have an entire hour to look around." Wrong!

I walked up to the window, stooped down, and in my very best Russian asked the dozing babushka on the other side if I could please buy a ticket.

"No," she answered without even opening her eyes.

"Why?"

"Because it is closed."

"But the sign says that it is open until four thirty, and it is only three thirty now."

"It is winter."

"Yes, I know, but the sign says you are open."

"In the winter, we are closed now because soon it will be dark."

And with that, she slammed the shutter closed on the ticket window and left me peering through the locked gates.

"Well, next week I won't eat lunch first. I'll come directly here after church," I muttered to myself as I headed, with typical Russian resignation, back to the train station.

I was most proud of the expedition during which I found, and purchased, a canary. I wanted a bird because it would be a little bit of company during the long winter nights. I mentioned this desire to my Russian-language teacher, and she informed me that on weekends there was a bird market in Moscow where all types of birds and cages were sold—plus everything else. She did not know just where it was but gave me vague directions. The next Sunday, I took the metro and then the bus to where my instinct said that a bird market should be.

Tumbleweed

What a flea market! Everything imaginable was for sale. There were many canaries to choose from. For the equivalent of two dollars, I bought a lustily singing orange canary and a small cage. I then tucked the bird into my yellow bag and covered it with several warm layers of clothing. Every so often on the long trek back to Star City, I would peek in to make sure it had not frozen or suffocated. From that night on, as I stomped the snow off my boots, I was welcomed home with a song. No matter how frustrating the day might have been, I smiled at this evidence of my basic Russian survival skills.

The most important of these skills as I explored Moscow on Sunday afternoons was discovering the location of the nearest public restrooms. Public restrooms were not plentiful in Russia like they were in the United States. I figured this was because Russians did not drink much water. There were no public water fountains either. If you didn't drink water, you didn't need restrooms!

While exploring on Sunday afternoons, I would purchase the items that I needed for the next week and could not find in Star City. For this purpose, I always had a large yellow canvas bag—actually, it was the same bag that one of my daughters had used all throughout high school to tote her books back and forth—folded up in my backpack. After I had stopped at the market to purchase fresh fruit, I would unfold the bag and fill it with my wares. When it bulged to capacity, I knew it was time to stop spending rubles. After all, I still had a long time to lug it around. Yellow bag filled, I headed back to the train station. There I caught the train to Star City. My goal was to get on the train early because it was always packed when headed outbound from Moscow, and I was partial to having a seat!

In the dim evening light, I would try to read, or I would doze, but I was careful to watch the time. After an hour, I started to look out the window so that I would know where I was. I always sat on the same side so that I could look out and see my "navigational" landmark—the jet trainer that was mounted on a pedestal in memory of a Russian test pilot. When I saw the jet, I knew it was time to count the stops to Star City. Finally, the train would stop at the Star City platform, and I would trudge off with the rest of the returning folks. We would go down the broken stairs, then wait for the train to depart so we could cross the track and head down the path that led into the gates of Star City. I loved returning to the peace and serenity of this wooded walk after the noise, traffic, and smog of Moscow. I walked contentedly, inhaling the

Lucid

fresh, crisp air deeply into my lungs, in no rush to enter the gates to begin another week of training.

Instead of Studying

On some Saturdays in spring and summer, to study nonstop was too much to ask of myself. So if I didn't have a test scheduled for the upcoming week, I would not listen to my mature self, exhorting me to study Russian. Instead, with the feeling of a kid playing hooky, I would hop on an outbound train, headed off to explore the various cities within a day's train trip of Moscow.

The first step for one of these mini adventures was to locate a destination on my map. I loved my map. I had found it while browsing through the many tables displaying items for sale that filled the busy Moscow streets. I could never pass up any of these tables, especially the ones with Russian books. Every time I was in the city, I purchased at least one book from such a table. I especially loved sounding out the author's name and discovering it was one I was familiar with and had read. Someday, I hope to have the time and ability to read the Russian books that fill the shelves of my study!

Back to the map. This map displayed all the train lines leaving Moscow. Using it, I would locate the city I wished to visit, determine which train line it was on, and figure out which train station in Moscow it departed from. Then, I would go to that train station, study the prominent billboard listing the schedules of all departing and arriving trains, figure out which track the train I wanted would be on, hop on the waiting train, and see where I ended up.

My favorite trip was to Zagorsk. The first time I traveled there by train, I arrived at the station I believed was the right one and disembarked. I had reached the end of my mapped route. I looked around at the throng of people and wondered, "OK, what's next? Where to now? How am I going to find the museum zone, the cathedral, and the monasteries?" Standing there on the train platform, I did a 360. Three-quarters of the way through my turn, I was temporarily blinded by flashes of sunlight being reflected off brilliant blue-and-gold domes sparkling under the bold blue sky. I set off hiking in that direction, with the golden domes as my beacon, and made it to the historic section of the city. Awed by the visual feast laid out before me when I stepped inside the walls, my first thought was that I would have to take Kawai here when she came for her Russian visit.

Tumbleweed

My daughter's first Saturday in Russia found me once again in Zagorsk. At lunchtime, on a park bench close to the seminary, we ate the sandwiches we had packed, watching the black-robed, gray-bearded Orthodox monks stroll purposefully past us, heads bent down. I was sitting in my normal position, legs crossed. A black-shawled grandmother, stooped from her years of toil, walked up and, uttering a string of angry Russian, tapped my leg. I did not understand her Russian, but there was no mistaking her tap. It was obvious that my crossed legs deeply offended her. We finished our sandwiches sitting bolt upright on our park bench.

As we left Zagorsk that day for our trek back to Star City, my daughter kept asking if I was sure that I knew how to get home. I assured her that it was no problem. Then she said, "Does NASA know that you just run around all over Russia by yourself like this?" Slightly taken aback by her comment, I realized my perspective had shifted—I thought of Russia not as a foreign country but as my home.

Some Saturday afternoons, instead of exploring Russian villages and landmarks by train, I took my bike and headed out on the trails through the forest. The forest outside the walls of Star City was a maze of well-trodden paths. I started randomly riding down them. I rapidly discovered that each was there for a purpose. All ended up somewhere. Some terminated at lovingly tended garden plots deep in the forest, but most came to an end at some small village. These villages usually consisted of a few weather-beaten wooden dachas, a well, and an assortment of free-roaming chickens, goats, a few milk cows, and children. Without exception, each had a monument erected at its center, built to commemorate the villagers who had died in World War II. Such monuments typically consisted of a stone needle three to four feet high, standing on a stone base where people placed wreaths. On the upright part, the names of the dead soldiers were inscribed. One afternoon, leaning against my bike, I began sounding out the names aloud. Most of the family names were the same. I looked at the long list of names, then around the small village. Every single male, every single father and son, from this tiny village must have been killed in the war.

Traveling by Train

On the map, Moscow appeared to be the hub of a gigantic rimless wheel, with the tracks of the electric train system being the spokes. I spent hours studying my map and plotting trips up and down these spokes. Little towns, basically just tall apartment buildings with an assortment of grocery stores, were located all along the tracks. Star City was one such complex. In between the different towns were huge, pie-slice-shaped uninhabited regions of dark and foreboding northern forests.

The trains relentlessly traveling these tracks were old. Many windows were broken, so in the winter the arctic wind whistled in, creating a bitterly cold environment. I will always remember the tears coursing down my daughter Shani's cheeks on her first train trip into Moscow because her feet were colder than she had ever dreamed feet could be. In the summer, it was just the opposite. There was no air-conditioning, of course, and if a few inches of space could have been found in a train car crammed with people and pet dogs, the proverbial egg could have been fried. No matter what the season, there was little conversation among the passengers.

On the plus side, the trains always arrived and departed on time every twenty to thirty minutes, rain or shine, snow or sleet. I never bothered with a train schedule. I just showed up at a platform and waited a few minutes, and a train would appear, headed in the direction that I wanted to go. All these train tracks converged at the Moscow stations, disgorging thousands of people all day long into the city. The train ride from Star City took an hour and ten minutes. From the downtown station, I could walk half a block, step onto the subway, and all of Moscow literally lay before me, awaiting my exploring feet.

The women of Moscow were always very dressed up, looking for all the world like they were headed to the small Oklahoma church of my childhood. To my amazement, they constantly wore heels—even while navigating the icy streets and paths in winter and the mud-filled cracks and holes at other times of the year. They were confidently treading paths in their heels that I would only try with great trepidation in my expedition boots! Next to them, I looked a little underdressed in my usual "astronaut casual"—tennis shoes, jeans, and a "worn in space" Lands' End knit shirt. The men were another story. They were invariably scruffy looking, with day-old beards and rumpled and soiled pants and shirts, looking like they had just been thrown out of a local bar. Silently observing these two disparate groups, many times the thought went through my mind, "Just who are these women dressing up for? Certainly not these men!"

Tumbleweed

To determine which train was the one headed back to Star City from Moscow, I had to consult the big outdoor board, which listed the outbound destination for each one by the number of the track. The only tricky thing here was that I had to read the Cyrillic carefully because there were two destinations—one being the Star City stop—that differed by only one Cyrillic letter, a sure pitfall for English readers who might want to skim the names too quickly. After finding the listed time and track number, I just went to the train and got on, always double-checking the destination displayed above the engineer's window.

The train leaving Moscow always started without any warning. Departure time arrived, and the train departed. That was it. Once when my son was visiting and we had spent the day in Moscow, we arrived at the train station. For some reason, he was in a rush to get back to Star City. We looked at the schedule, and he shouted, "Our train is leaving in three minutes! Hurry! We can make it!" He took off running for the track.

"No, we can't make it, Michael. Stop!" I yelled as I chased after him, encumbered by my heavy backpack and my bulging yellow bag loaded with produce.

He got to the train, grabbed a handhold, and started to swing up. He got one foot on the platform before the door slammed shut on his foot. There he was, an over-six-foot-tall teenager hanging on to the train as it started to move, with one foot caught between the closed doors and the rest of his body hanging out. I yelled at him to get off even though I had no idea how he would manage it with his foot trapped. Two Russian men inside the train saw his predicament, and with brute force, they pushed the doors open and shoved Michael's foot out. He plunked down on the platform as the train accelerated out of the station. I ran up to him and, after making sure all his bones were in one piece, asked him just what he thought he was doing. What did he think he would do if he had ended up on that train and left me behind? How would he have known where to get off? What would he have done upon getting off the train in some strange Russian village without knowing where he was or even where he was going? After I finished my tirade, he just shrugged and said, "Mom, if you had run just a little faster, we could have made it."

Sometimes a conductor would come through the train car and ask for your ticket, but most of the time, no one asked to see it. I explained

Lucid

the system to Michael as we were getting ready to go into Moscow by train for his first time. I gave him his ticket and said, "Now, be sure not to lose it." When I asked him if he wanted me to hold on to it, he assured me that he was an adult and could handle his own ticket. "Don't lose it," I admonished as he placed it somewhere in the multitude of his winter garments, layered for warmth.

The train was packed. I was reading my book. Suddenly, a conductor was standing by me, asking for a ticket. I got mine out and he punched it. He stretched out his hand to Michael. "He wants your ticket," I said, pulling him out of some deep reverie. Startled, Michael began to hunt, first in his outside pockets and then reaching in deeper. He stood up. People on all sides leaned away from his jutting and flailing arms as they went into inside pockets and then outside pockets, this crevice and that. All to no avail. The conductor was getting impatient. He said something.

"What does he want?" Michael asked.

"He wants you to pay a fine. You owe him ten thousand rubles because you do not have a ticket."

"Will you pay?"

"No way. You were responsible for your own ticket! You have money. You pay."

Michel found the ten thousand rubles, and the conductor wrote a ticket. Later, Michael added it to his collection of traffic violations, saying that he had started a new hobby—collecting tickets from around the world.

My system for finding the Star City stop (simply counting the stops past the plane replica) worked great. Only once did I run into a problem. It was a hot, steamy summer day. My husband and I had spent the afternoon walking around Moscow. Arriving at the station, Mike was good and cross. We got on the train. It was jam-packed with bodies. I got out my book and started to read. There was an unintelligible announcement on the scratchy loudspeaker. Immediately, half of the people poured off the train. I shrugged. More room to stretch out. The train started. I dozed off. Later, I woke up with a start. I glanced at my watch—outbound for an hour. I looked out the window. It was not the normal scene. I sounded out the Cyrillic name at the next stop. Never heard of it.

Mike looked at me and said, "Do you know where we are? Are we lost? Are we on the wrong train?"

Tumbleweed

"No, we are not lost, and yes, this train is not going to Star City."

"What will we do?"

"Don't worry. We'll get off at the next stop and take the next inbound train to Mytishchi, a stop where you can transfer trains without going all the way into Moscow. Then we'll catch an outbound train for Star City."

"How do you know that will work?"

"Just trust me. I understand the Russian train systems."

We got off at the next stop. I went into the station and bought tickets to Mytishchi. There we were, a middle-aged American couple, sitting beside a train track in an unknown Russian village in the gathering twilight. Mike was about as grumpy as he could get. He was not happy being in a small town of which he knew neither the name nor location, being forced to have faith that his wife knew how to get him back to his bed for the night. Sitting there, waiting for a train that I was sure would sometime come, watching a young girl herd her ten cows across the track as she headed home for the evening, listening to the low rumble of Russian being spoken by the villagers also hanging out beside the tracks, not knowing exactly where we were—this made my list of top ten "mystical Russian moments."

Christmas

John and I were not given any time off to go home for Christmas, so our families came to Star City to be with us. Because of my family's varying work and school schedules, everyone arrived at different times. Michael came first. I could not meet him because of my training schedule. Charlie Precourt, then the NASA DOR, offered to pick up Michael when he picked up his own family, also coming in for Christmas. Their planes were to land at about the same time. I described Michael to Charlie and Charlie to Michael. Michael's plane was late, and his baggage was missing. It took a long time for him to get through customs. Somehow, he missed Charlie. He wandered around the airport, then went outside and spotted the NASA van. Charlie's daughters were inside. When they saw Michael, they hopped out to find their father. Michael followed them. Four American teenagers wandering in separate directions in the dimly lit Moscow airport was not the best way to begin a Christmas vacation. Charlie was definitely not in control at that point, but somehow, he did manage to

Lucid

get them all into the van and headed out to Star City. After hearing the story, I realized that I would be forever in Charlie's debt.

Michael wanted to go to the Moscow Zoo. We had tried to do that when he visited in May, but the lines were too long. On Sunday morning, with the temperature way below zero and the skies spitting snow, we arrived at the zoo. There were no lines. I didn't think it was even open, but Michael looked around and found a tucked-away booth where an older woman was sleeping with her mouth wide open. I rapped on the window and asked if we could purchase tickets. We bought two and went in. We were the only ones there. The polar bear mesmerized us for a large portion of the morning, performing toss-the-Frisbee for us. Never in my life had I ever seen such an active polar bear.

Later, Michael suggested that we find the natural history museum. For several hours, we roamed up and down the winding, snowy Moscow streets, wiping snow out of our eyes and looking at our wet map. Finally, I saw a small plaque on the side of a building and realized we had arrived. We went in, deposited our coats, and spent several hours sauntering up and down halls lined with bottled specimens. As usual, Michael had to find a restroom. His nose led him under the belly of a mastodon skeleton that filled a dark corner of a stairwell and into the men's room. Upon emerging, he commented that it was a good thing his sisters were not with us. The bathroom he had just used was below even his standards.

We eventually decided it was time to leave if we were going to get back to Star City before dark. Stepping out into the Russian afternoon, we entered one of those all-too-rare mystical fairy-tale moments. The snow was swirling large, lacy flakes. A small slit in the overcast sky allowed a shaft of sunlight to beam through and light up the golden domes of the Kremlin. Michael and I gasped in delight and then plunged homeward in the biting cold of the gathering darkness.

The rest of the family finally arrived, but among them, only Kawai had come with a specific list of things to do and see while in Russia. Naturally, that became our agenda. At the top of her list was a visit to Tolstoy's summer home. The adjective *summer* should have given us a clue as to the suitability of this destination as a midwinter excursion. I arranged for a NASA van to drive us. I was assured that the driver would know how to get there. In the early-morning darkness, the six of us, plus a Russian driver, piled into the van and headed out of Moscow.

An hour outside Moscow, Michael said, "Mom, have the driver pull over. I need to go to the bathroom." Looking out the window, there was nothing but open road and fields of snow. Not a bush in sight. I ignored my son. His clamoring became more insistent. I told him that this wasn't Houston, with a gas station on every corner. He would just have to wait. But when Michael threatened to jump out of the van, I mentioned his plight to the driver. At the next military checkpoint, the driver pulled off the road and pointed to a dilapidated outhouse leaning behind it. Michael jumped out. Sensing an adventure, Shani's husband, Jeff, grabbed a camera and jumped out after him. The driver spotted the camera and shouted, "No! No! Not allowed! Military checkpoint!" and pursued Jeff, grabbing his camera as they both tumbled into a snowbank.

After what seemed like an endless stretch of hours and police checkpoints, we finally drove up to a gate at the end of a narrow road. "We are here," the driver announced. "I can drive no farther."

We piled out. No one was in sight. I went up to the gate and saw a small window in a guardhouse. Through the foggy glass, I could see two young women and a computer. I rapped gently on the window with my gloved hand. The window cracked open. I asked if we could buy tickets to see Tolstoy's home. The young women looked at each other, slammed down the window, and rapidly conversed with one another. Then the window cracked open again.

"Where are you from?" one of them asked.

"From Moscow," I replied.

"No," the woman said, then slammed down the window again.

We stood and waited, every member of the family espousing their own favorite theory about what was happening. This scenario repeated itself twice. Then Kawai said, "Mom, they don't want to know where you came from this morning but what country you're from. They know you aren't Russian."

The next time the "where are you from" floated out of the crack in the window, I answered, "Houston, Texas. In the United States."

That answer generated action. Fingers flew over the computer keyboard. Several pages were printed out, and then I was asked for rubles. I handed over the rubles, the gate swung open, and we walked up a long driveway. No one was in sight.

Lucid

Finally, we arrived at a white frame house. We walked onto the porch. No one appeared. The doors were locked. I sat in a rocking chair and listened patiently to my family complaining about the cold, about Kawai's choice of destination, about not knowing what was going on.

Eventually, a door opened, and we were invited in. A young girl who had obviously just been hustled up from the nearby village invited us, in broken English, to tour the house, with her as a guide. The house was frigid, a fact Jeff and Michael complained about in every room. Our guide was both informative and enthusiastic, filling our cold brains to overflowing with details of daily life during Tolstoy's time. But even I thought that the word *summer* was mentioned just a few too many times.

Once the tour finished, we were ushered out, and the door was locked behind us. Kawai insisted that we find Tolstoy's grave before going back to Moscow. We saw a sign and slogged through the snow in the direction of the arrow. We found the grave under towering fir trees and took a family portrait. On the way back to the driveway that would bring us to the van, I saw what I was sure were restroom facilities. We went in and discovered the highlight of Tolstoy's summer home—a heated outhouse!

In the darkness of late afternoon, we headed for home through the Moscow suburbs, with overloaded trucks barreling past us on every side, trying not to breathe the cold, exhaust-laden air too deeply. Suddenly, there was a sharp crunching sound. The next thing we knew, we were stopped on the side of the highway along with the truck that had sideswiped us. Both drivers jumped out and began to berate each other in increasingly loud Russian. Michael kept asking, "Mom, what are they saying? What are they saying? Get out there and tell us what is going on." With as much sternness as I could muster on that frozen night, I turned to Michael and told him to be quiet and not to interfere.

The trip to Tolstoy's summer home quickly escalated to epic proportions in the Lucid family myths. Now, whenever someone starts to complain about something, the retort is always, "At least we aren't at Tolstoy's summer home." And yes, Kawai was demoted from her position as the Lucid family travel manager that Christmas!

In Russia, Christmas is not celebrated on December 25, so that date is a working day in Star City. John and I both said that we would not work that day but planned to spend it at home with our visiting families. For weeks, this point was relentlessly argued. The Russians declared that there would be no time off on December 25. We could have the Russian Christmas as a holiday instead. We tried to explain that our families would be with us on the American Christmas, not the Russian Christmas.

Finally, John and I announced that it didn't matter what would be scheduled on that day—we would not show up. We figured that if NASA wanted to fire us for taking such a stand, so be it.

We did not work on Christmas Day. Instead, I was with my family. We were in my apartment, huddled under blankets, trying to survive the worst cold of the season. The week before, I had been able to pick up a turkey at the American Embassy, and I was trying to cook it. The oven in the apartment did not want to heat up. Finally, we decided to eat our festive meal in segmented courses. First the dressing, then the mashed potatoes, and, a few hours later, a semicooked turkey. We ended the holiday by going hiking outside in a blizzard, which proved to be somewhat warmer than the apartment.

We said our goodbyes, and my family returned to Houston while I remained in Star City for the final training sprint before heading off to Mir. I gave thanks for having such a tremendous family that was willing to fully support me in this adventure. As I marveled at the uniqueness of each family member and how all contributed to the strength of the family structure, I reread the letter my daughter had sent me after her summer trip to Russia, which she had titled, "Window or Aisle?"

Not so long ago, I saw a television program about surviving a plane crash. Contrary to popular belief, most people in a crash don't die on impact; they die of smoke inhalation as they jam the aisle to escape the inferno that had once been their comfortable (no smoking allowed) plane. The trick to surviving a crash is planning ahead, which simply involves choosing an aisle seat within three rows of an exit. Sit anywhere else and your survival odds literally go up in smoke.

Always having been enamored with safety (in a world where it's harder to find each day), I've taken to sitting in aisle seats, always within three rows of an exit. The only airline I fly much at all is Southwest, and as long as I'm willing to sit in the middle or back, it's never hard to find a seat that meets my criteria. All I have to do is arrive a little early. Since none of the flights allow smoking, I don't even have to worry that sitting in the back will give me lung cancer. All I'm really giving up is the chance to be first off the plane. Not sitting by the window isn't a concern either. I've been everywhere I go on Southwest and have already seen the view.

Lucid

I'm on my way home, though, from traveling to Russia to visit my mother, who is working in Star City for a year. I wasn't given a choice of seats for the flight over. I had no regrets as I took the aisle seat (by chance, only two rows from the nearest exit) assigned to me. I crossed half of North America, the Atlantic Ocean, and much of Europe in secure comfort, content to watch my journey on the map constantly displayed on the overhead monitor. When I landed, I stepped foot on Russian soil without ever having seen it from the sky.

Returning home, my experience has been altogether different. After surviving the mandatory hour wait in the mob trying to go through Russian customs before leaving the country, I finally made it to the ticket counter. The Russian KLM employee cheerfully asked me which seat I preferred, window or aisle? He smiled and waited for what he surely expected to be a quick response. How could he know that in the guise of a simple question, he had touched upon one of the central dilemmas of my life? Be safe or take a risk? Be safe or see the view?

Certainly, I had not escaped the question during my two weeks in Russia with my mom. How could I? From the day my mother was four years old and set out on foot to journey around the world, she has always taken a window seat on the train of life. It is a trait I admire in her, even as I urge her to be cautious. It is not a trait I share. From my seat on the aisle, I strain to catch a glimpse of the landscape. It is difficult to see anything clearly in the images hurtling by, and I often turn my attention back inside to count the number of rows there are to the nearest exit, just in case the worst happens and we have to feel our way out as we escape the crisis through the blinding smoke. It is an obvious difference between me and my mother. I step cautiously, while she rushes forth. No wonder we often have a hard time walking side by side.

As we explored Moscow (where, thankfully, Mom had scouted the territory before I went with her), we frequently traveled by subway. Mom loitered in each of the stations, pointing out the tiled mosaics on the wall of one stop, the stained glass of another. I dutifully looked where she pointed, then returned my attention to the people around me. I watched their movements and held my valuables close. Only after I'd become comfortable with the routine of subway travel (which, admittedly, wasn't until my stay was just about over) did I pause to look more closely at the unique architecture of each station. Even then, I did not stop for long.

Of course, being in Russia with Mom, it was only natural that we spend some time riding bikes. On the walled streets of Star City, we rode side by side, but as we entered a rural village away from the complex, Mom quickly pulled ahead. Not knowing Russian, I could not read the signs. Not knowing Russia, I didn't know the expectations governing who travels where. I lagged behind, insecure in my conspicuousness. Mom led us through a gate and into a field. Pumping hard to

Tumbleweed

make it through the mud, I passed close enough by a cow to touch her nose. We also pedaled, ever so slowly, by an old man trudging along in the opposite direction. He was weary and worn and none too clean, but I thought he must be glad to belong.

My second Saturday in Russia, Mom and I rose early to catch the train to Zagorsk, the longest continually operating monastery in Russia. The Moscow guidebook suggests visiting with a tour group. Train travel is recommended only for the adventurous. Mom certainly enjoyed the ride. I crammed myself into the corner of my seat and tried to be inconspicuous, hoping no one would speak to me. If my appearance had not already betrayed my foreignness, my speech certainly would. Without language, even a world full of people is empty and none too safe.

Once in Zagorsk, I sighed in relief as I caught a glimpse of the tour buses in the monastery's parking lot. To me, the sun reflecting off the silver tops of the buses equaled the beauty of the gold-domed cathedrals. The buses assured me that it was permissible for me, an outsider, to be in this place. Even as my own comfort level increased, I recognized in my mom a wish for something more. Her dreams of adventure rarely include landscapes cluttered with tour buses.

Window or aisle? The KLM employee got his answer. Maybe the realization that I had traveled in a Russia that an outsider would not have been allowed to see only a few years ago enabled me to take a little piece of the sturdy Russian soul for myself. Or maybe the small part of me that is my mother called my imagination to action and overpowered the cautious part of my brain. I pondered the rationale for my choice as I boarded the plane and took my seat by the window.

I sit now gazing out the window at the sea of clouds below me. Not having landed yet, I don't know if I am safe. The logical part of my brain tells me that statistics are against anything bad happening to me during the flight. The rest of me wants to double-check the safety information in the seat pocket in front of me when we hit a little turbulence.

Safe or not, I have seen the bleakness of the Moscow skyline fade into the lush Russian forests. The waves of the Baltic Sea have rolled beneath me, carrying a tiny ship to—where? Perhaps to Finland, whose shores I have looked toward from land, but have finally seen from the air. I have seen the fields of Denmark, and even caught a glimpse of German soil. Soon, we will reach the shores of Holland and I will have a good view of Amsterdam's canals. As we prepare to land and the ground rises up to greet us, I may also see a windmill or a field of tulips. Although I won't know it, I may gaze upon a place where my great-grandparents once walked.

Safe? Probably more so than my imagination would lead me to believe. (And I do take comfort knowing that only one small man sits between me and the aisle, and that I am only three rows from the nearest exit.)

Lucid

Safe? I see Holland below me and know that for some deep part of me, a part of me I only see reflected in the color of my eyes and hair and occasionally in my grandmother's dreams, for that part of me, I am seeing home.

Window or aisle? One answer isn't always the right one, but this time, I have chosen well.

—Kawai Lucid
August 18, 1995

TRAINING

Whenever I talk to anyone at NASA about training for my flight on Mir, they immediately envision a process very similar to NASA shuttle training, just replacing the shuttle with Mir and Soyuz, and oh yes, a little strange-sounding nomenclature and a few Cyrillic letters. Nothing could have been further from the truth.

After agreeing to undergo training for my Mir mission, I began to wonder what I might have gotten myself into. I thought back to my last trip into space, SLS-2, a Spacelab mission dedicated to life-science experiments. I had been assigned a year and a half prior to launch. Because it was my fourth shuttle flight, I was already familiar with my shuttle duties, so I spent the bulk of that year and a half training on just the life-science experiments.

In comparison, for this proposed flight on Mir, I would have just a year in which to train on two totally new vehicles—Mir, where I would be living, and the Soyuz, our lifeboat while on Mir—and, in addition, learn all the NASA scientific experiments. Plus, I would have to retain familiarity with the shuttle because I would be launching and returning in it. Of course, all this training would happen not only in another language but one with a different alphabet. I gulped and wondered again what I had agreed to.

Learning the Language

What I did not realize then, and what most people not directly involved do not understand, was that the Russian concept of training totally differed from NASA's conception. The two ideas were as dissimilar as

Lucid

the Congo is from the Arctic. Imagine how a Pygmy dropped into an Eskimo village and told to hunt whales or an Eskimo dropped into the Congo and told to hunt elephants might feel, and you will begin to understand how John and I felt when we started our training.

The superficial differences confronted me within an hour of arriving in Star City when I was handed the schedule for the next week. It was all in Russian, the entire sheet swimming in Cyrillic letters that didn't seem to join together to form any of the words in my limited Russian vocabulary. That first evening, as soon as I was alone, I got out my dictionary to puzzle out what life now had in store for me.

Perhaps it was a good thing I did not yet realize that I would work harder during the coming year of training than I had ever worked in my life. I certainly worked a lot harder than when I was in graduate school with two toddlers at home! What made the year so extraordinarily difficult was that we lived and worked entirely in Russian. Many weeks, the only English I heard was in the conversations John and I had as we walked together to and from class. It was like trying to live and work using a pair of glasses with the wrong prescription. Everything you see is fuzzy and out of focus. The constant straining to figure out what exactly it is that you see is exhausting. Wearing such a pair of glasses, you strain all your eye muscles, your face muscles, your neck muscles, all the muscles in your body just trying to bring the world into focus, trying to get around, and it just does not work. The world remains out of focus. And it literally wears a person out. And that is what it was like to work on daily training in Star City. In addition, because of my lack of vocabulary and inability to put the language together, I felt like only a partial person.

Before starting Russian-language training, my foreign-language experience had been minimal. I had taken German in college to meet the language requirements for my chemistry major, and during the first two weeks of my freshman year, I knew I was in serious trouble and went to the teacher for help. When she asked me what my problem was, I said, "Well, you keep talking about these things called nouns; now, just what are you talking about?"

She stared at me for a long moment and then said, "You do have a problem, and I don't know how to help you." Fortunately, we were able to get extra-credit points in the class for memorizing lines of poetry, so I would sit in my dorm room, memorize a couple of lines of German poetry, and run to the language lab. There, I quickly scribbled

Tumbleweed

them down. As soon as these lines flowed onto the paper, a blank space was left in my head. I quickly ran back to the dorm and filled this blank space with a few more lines of German poetry and repeated the entire process. And that is how I passed German. Over the years, the story took on mythic proportions in our family lore, becoming a "Lucid Legend." So it was totally understandable that when I announced to my family that I was off to language school to learn Russian, the laughter was so long and so hard that we could not finish supper.

Our first six weeks of training in Star City consisted entirely of Russian-language instruction with two teachers who, incidentally, never spoke any English to us. They worked with us four to six hours every day on the technical Russian we needed to understand before we could begin our Soyuz and Mir systems classes. During those first few weeks, we were given two hours every morning, from nine to eleven, for self-study. Seeing these scheduled two hours always made us laugh because our mere survival in the program seemed to require spending almost every waking moment on self-study. I woke up at five thirty every morning and quickly dressed and ate a piece of toast. Then, I studied Russian until it was time to go to class.

Typically, John and I met our Russian-language teachers, who commuted daily from Moscow out to Star City, in the tearoom each morning. A few cosmonauts and the two French cosmonauts—the prime and the backup, also training for a flight on Mir—gathered daily for a tea break. As we sat in the red-and-black booths drinking our tea, we tried to converse with everyone in our halting Russian. After ten to fifteen minutes of this, we were off to class with our teachers.

We met in the same big classroom each day. We didn't turn on the lights because they generated such a loud background hum that we couldn't hear the teacher. Because the majority of those days in January and February were overcast and snowy, we spent most of those first six weeks sitting in the semidarkness trying to understand technical Russian.

First, we learned the general construction of technical Russian sentences. Then we read and memorized paragraphs about the Soyuz or Mir. Many times, we had to write down these paragraphs verbatim when we got to class.

John and I both love to talk, so we felt a lot of pressure to learn enough Russian to be able to converse with our Russian crewmates

Lucid

once aboard Mir. Unfortunately, our learning styles were totally opposite. John wanted to hear the new words and phrases, whereas I had to see them written on the board before I could repeat the new sounds. This was a source of frustration for the instructors as well as for us. Finally, toward the end of our time in Star City, John and I convinced the teachers to teach us individually. This gave us one-on-one time with the teacher and helped decrease our frustration levels.

Although training in the language to be used aboard Mir had certain advantages, there were a number of disadvantages to a training program conducted entirely in a language in which a person is not proficient. I realized this most clearly once I had returned home from my stay on Mir. I was back at the Johnson Space Center, sitting in a meeting in which we were talking about the recent fire on Mir. I was asked if I had ever been trained on the use of a Mir fire extinguisher. I said yes, I had seen one, lying on a table. The NASA manager running the meeting asked if I had ever activated the fire extinguisher during the training. I replied that we had not because there was not enough money in the Russian system to provide actual working hardware. The manager then commented that maybe that was the difference between a pilot (himself) and a scientist (me) because a pilot would have insisted on handling the hardware. What he failed to realize was that the entire time the fire extinguisher was on the table, John and I were trying to figure out how to pronounce and write the Russian word for it. We were focused on the vocabulary that would be on the next day's oral exam. We didn't ask to handle the fire extinguisher because the Russian system of training focused on our knowing the vocabulary to pass the exam, not on "hands-on" practical experience.

During this time, I thought a lot about language and what language means to an individual. I have always loved words. I have always loved to read. I think the hardest part of working in another language was the handicap inflicted on me by the lack of a large vocabulary. We humans truly are the words we know, and this was brought home daily to me while working in Star City. Many times, I felt that I had been stripped of my personality because without words and all the nuances of words, I could not express my real self. I felt that when someone looked at me, all they saw was a not-quite-bright American woman. I felt handicapped and frequently imagined that everyone listening to me speak thought that I was just a tad bit "touched in the head" because

Tumbleweed

I could never really say what I was thinking. They had to play some sort of guessing game to figure out what I was trying to communicate.

At the beginning, I felt good about my ability to speak Russian. I was full of confidence. I tended to forget that it is always easy to feel confident of an untried ability. When our schedule showed that we were going to spend a morning in a Russian airplane flying parabolas, I was excited. (The purpose of parabolic flight is to produce about 30 seconds of microgravity at the apex of each parabola.) It would be one of the first times that we were going to be with a couple of the new Russian cosmonauts who we had heard were training in Star City. I was especially interested in meeting Nadezhda Kuzhelnaya, the one Russian female who was in training to become a cosmonaut. It would be a great opportunity to try out my Russian skills on someone other than our teachers.

The big day came. We were in the airplane. The pilots were flying the parabola profile. At the top of each parabola, we floated in the belly of the plane and then grabbed the lines tied along the sides as the aircraft dipped to the bottom of the parabola. At the bottom, we were pinned to the side of the plane as it began a 2 g pull-up in preparation for the next thirty seconds of 0 g. I found myself next to Nadezhda, so I started talking to her. I described our training and the exercise we were currently doing and asked her questions about her background. She smiled and listened to me with rapt attention. My confidence in my linguistic abilities soared to great heights. And then, over the roar of the engines, she yelled in my ear in broken English, "What language are you speaking?" And there went my confidence!

My struggles with the Russian language provided me with an unexpected gift: I now have an entirely new appreciation of language and what it means for a person to be robbed of it. Language is precious, and a person lacking a rich vocabulary with which to express him- or herself is truly handicapped.

Useful Language

Although I spent hours each day learning Russian, it took only a short walk through the streets of Moscow to realize that I wasn't learning the "everyday" language. Most Russians had no interest in talking with me about ways of producing oxygen on Mir, which, incidentally, due to the Herculean efforts of my Russian teachers, I could do very well. When I was stopped in the street and asked directions or when I was trying to buy something in one of the city's many kiosks, I found that I had huge gaps in my vocabulary.

To rectify this problem and to make sure that, during our stays on Mir, we could do more than discuss how the waste management system worked, John and I asked a teacher at the Star City high school to come over to our apartments several evenings a week for tutoring in "real" Russian, the Russian that the everyday people spoke during their daily lives. Because NASA would not pay for these lessons, John and I paid for them ourselves.

Three times a week, Nina would appear at my door, in her best dress and, no matter how much snow there was on the ground, high heels. I met her at the door in slippers and my warmest sweat suit. For one or two hours, we would talk. She would tell me about the daily lives of the Star City residents and ask questions about life in Texas and my family. She told me about her life. After she had earned her teaching degree, she traveled to Eastern Russia to teach school in a small village dominated by a coal mine. There she met and planned to marry a young man. There was a fatal accident in the mine. Twelve of the thirteen bodies were recovered. Only the body of her fiancé was not found. She described the funeral procession winding through the muddy streets of the village with thirteen coffins—one of them empty. She never married and had taught school ever since.

As my Russian vocabulary increased, she asked me more and more questions. She was very intrigued by the fact that I went by train every Sunday to church. She wanted to know if she could go with me. I assured her that would be a wonderful idea, and then she asked me to describe what I saw, what I looked at in church. I said, "Well, it is a movie theater, and there is just a faded brown curtain across the stage."

"Please tell me what there is on the walls, what the icons look like," she said.

"There is nothing on the walls; there is nothing to look at," I replied.

"If there is nothing to see, then why do you go to church?"

My answers proved unsatisfactory, and she decided not to go to church with me because there was nothing to see.

Many times, our conversations drifted to the topic of God and what people believe. She asked me once if I knew anything about "God's laws." When she was a small girl, her aunt had told her there were laws of God, and she had wanted to know about them ever since. I explained the Ten Commandments to her. I explained further that Jesus had taught that the greatest commandment of all was "to love the Lord your God with all your heart, with all your soul, with all your strength, and with all your mind and to love your neighbor as yourself." She was entranced with the concept.

"That is just what my students in high school need to know about! I am going to copy them from your Russian Bible and hang them in my room so that the students can read them every day," she exclaimed.

Ironically, Nina, living in a country only a few years removed from communism, felt free to hang the Ten Commandments on her classroom wall, but such a thing would never have been allowed in the classrooms of any of my own children, who had attended public school in the "land of the free."

The Real Thing
After we had been in Star City for about six weeks, the military officer in charge of our schedule marched into our Russian-language classroom. He announced to our teacher that we would begin systems training the following week. Therefore, our time in Russian-language study would be decreased. In the weak winter light, I actually saw her blanch.

"But they are not ready yet," she stammered.

"No matter," he replied. "Time is short, and there is much they need to know."

Then, he marched out.

We entered a new phase of training. Gone were our treasured self-study hours in the morning. Instead, we joined the throng of Star City residents entering the working section of the city at nine each morning. Our days became a constant stream of two-hour instruction blocks, from nine in the morning until six in the evening, with a one-hour break for lunch. We really grew to appreciate the instructors who were addicted to nicotine because they would give us a five-minute break midway through the two-hour class block!

Our classes started on time and they ended on time. They never ended a few minutes early, and they never ran over, except once. It was early in our training, and we did not yet realize the depth of the cultural difference between the Russian method of training and the American one. Unlike the American system, which had built-in flexibility to it, we discovered that the Russian approach was dominated by rigidity.

Lucid

Typically, when you took a class or participated in a training exercise for the shuttle, about half of the time was spent on questions and answers between the instructor and the student. If it took longer than planned, it wasn't a problem. At the discretion of the instructor or the crew, a makeup session would be scheduled. On this particular day, we were having a lecture on a Soyuz system. During the lecture, we frequently asked questions in our halting Russian. The instructor became visibly agitated. At the end of the two hours, he said that he was not finished with the lecture. We replied that it wasn't a problem and that we'd gladly return at the end of the day to finish. He replied that we did not have permission to do that. We insisted, saying it was no problem. He told us to wait. He left and swiftly returned with his boss, a higher-ranking military official. He told us it was not possible to come back. Then another military official came into the room to find out what the problem was. It was obvious that our instructor was uncomfortable, so we just said that we were satisfied with the lecture and did not need any further instruction. After that incident, we were careful about asking questions and made sure that any we asked did not make the lecture go over the allotted time.

The type of questions that we, as American astronauts, typically asked highlighted another cultural difference between the space programs. Usually in a shuttle system lecture, we would ask questions like, "Well, if X breaks, what happens next? How do you work around that failure?" In Star City, anytime we asked that type of question, we were met with blank stares. It was clear to us that we were facing something more gigantic than just a language gulf. After the initial blank stare, we always got one of two answers. The first answer was that the question did not make sense because nothing on Mir ever broke. Supposedly, everything on Mir was in the same pristine condition in which it had been launched. The other answer to our questions was, "It's broken? Well, replace it!"

After I had spent a little time on Mir, I finally understood that blank look. No, everything was not in pristine condition; things broke daily. However, the procedural work-arounds to correct malfunctions on the shuttle were not done on Mir. On Mir, a panel near the broken part would just be opened to reveal the spare part destined to replace it.

Many of our working days were eight solid hours of nonstop system lectures, all in Russian, of course. Talk about an "end of the day, my head is too full" type of headache! I developed a makeshift strategy to

Tumbleweed

deal with this. I would take notes. The instructors who won my undying affection were the ones who would print, in block letters on the scratchy blackboard, the Russian names of the equipment pieces they were describing. When that was not done, I would try very hard to scribble some type of English phonetic version of the sound that the instructor was making. Later that night, I would repeat what I had written, convert it into Cyrillic, and try to match it to some piece of the system puzzle. This was done using a lot of intuitive knowledge of how pieces of equipment work.

I rapidly discovered, much to my dismay, that the Russians depended much more on aural learning than visual learning. I have always learned by reading, not by listening, so the fact that there was a real paucity of written materials was difficult for me. John and I soon learned that many of the systems discussed during our lectures were also written up in what the Russians called Komspects. At the beginning of each lecture on a new system, we always asked for a copy of the Komspect describing the system. Often it took repeated requests before we were reluctantly given one copy, generally a barely legible reproduction. (Because these were all in Russian, the poor reproductive quality made things especially difficult for us.) We were made to understand that this was just a temporary loan and that we were required to return the Komspect in a few days. We would then try to find time to photocopy it at the NASA office and turn it back in. The next step was trying to translate it so that we could puzzle out the system we were currently studying.

Time and time again, I would spend all weekend going through one of these documents, only to have the feeling, "So what?" Of the Komspect's fifteen or twenty pages, there might be one or two paragraphs of useful information. This was another big cultural difference. English technical documents are terse and to the point. Most contain no filler words. Technical Russian is not like that. At times, it seemed that 90 percent of a Komspect was only flowery filler phrases in which a few relevant gems were embedded. Slogging through all this excess language made deciphering these documents incredibly frustrating. It was like a continual treasure hunt with vague clues and without the certainty of any treasure!

This was vividly illustrated one morning when John and I were separately scheduled for one of our very few sessions in the Mir mock-up to go over a procedure. I was in the mock-up first, along with an

Lucid

instructor. I was given a procedure. I stared at the lines and lines of Russian and started to wade my way through.

Then the instructor said, "All this is not needed," and drew a red line through all but one line of the procedure. "You just need to remove the cap. That is all."

So I did, extremely thankful that I did not need to translate all that Russian. After I finished, I went to our language classroom and found John there. He had spent the last forty-five minutes with the same procedure that I had just been shown, trying to translate it. I then drew the same line my instructor had drawn on mine and said the same thing: "All this is not needed! You are just going to remove a cap."

John looked in disbelief at the paragraphs of translation that he had done. We muttered together about a language that had to use one hundred words to indirectly command a simple "remove the cap" action.

Not only was there pressure to learn the Soyuz and Mir systems, but there was the additional burden of learning the technical vocabulary so we would be able to pass the oral exams on each one. After a lecture on a particular system, we would be scheduled for a "consultation" on it, and then we would have the oral exam. The consultation was conducted by the main instructor for the system, and he would review what, in his opinion, we needed to memorize for the oral exam. For instance, we would be told that we needed to explain what the functions of the system were, then the list of specifications, and so forth. Oddly, we were never told how the system was used— such as crew procedures or other practical, operational things. After the consultation, we spent hours studying.

On the day of the exam, John and I would show up together at the appointed time in the designated room. A panel of experts from the Russian space agency would be seated behind a row of tables. Our instructors would also be present. In addition, we always had a huge pictorial of the system we were being tested on and a pointer that we could use. One of the Russian space agency engineers would ask a question, which we were to answer. This was all done in Russian. There were never any translators present or any English words spoken.

The first hurdle in one of these exams was to figure out what, exactly, the question was. Many were the times we could not even fathom a guess. We would then just shrug our shoulders in unison, take a deep breath, and plunge ahead. First, John and I would discuss

Tumbleweed

what we thought the question was, and together we would come up with a collective guess. It always comforted me to realize that my darkness was being shared! Then, John and I would take turns trying to answer a question we never understood in the first place. When I didn't have a clue as to what the question was, I would simply take the pointer and point to parts of the system on the big diagram and just say the words that I had spent so long memorizing, throwing in a few Russian verbs and connecting words so that there was a semblance of a sentence. A couple of times at the end of the questioning, we were sent out of the room to stand in the hall to wait while the panel of experts deliberated the verdict as to whether we had passed or failed. Incidentally, I never did figure out what the criteria were. I assumed that we passed, because we did not ever repeat a class. Of course, in our inflexible schedules, there was no time for repeats.

After we had taken a couple of these tests, there was one thing that I knew for sure: whatever the instructor told us to be sure to know would not be on the test, and whatever he said we did not have to memorize would be on the test. Also, after a couple of tests, it became evident that the main purpose of the exams was not to test us but to see if the Star City instructors were doing their job. In other words, the Russian space agency that contracted the training out to Star City was using the excuse of testing us to check up on and intimidate the Star City training system. We, the non-Russian-speaking students, were the victims!

At the end of our "survival" consultation, I was so frustrated that I was ready to go back to my apartment, dig out my return ticket home, and just say forget it. For the last hour, the instructor had been telling us that we needed to memorize all the different calibers of bullets. Under the commander's seat in the Soyuz, there was a gun to be used to shoot bears or wolves if you made an emergency landing in Siberia. Why I, who had never even handled a gun before, had to know these kinds of details—and in Russian, no less—was beyond me. As John and I walked out of the building together, looking down to make sure we did not slip on the sheet of ice covering the steps—there were no handrails to hold on to—John assured me that he could explain to me in English what *caliber* meant. Then he repeated the lines that we said over and over again to whichever one of us needed it the most at that time: "What are they going to do if they don't like our answers? They can't send us back. There is no one else to take our places, and if we aren't here, the money isn't here! All that can happen is that they can make us sit through the lectures again." True words, but they could not really lessen our test anxiety.

Our funniest test experience happened toward the end of our training flow. We had received our schedule on Friday, as usual. There was no mention of any kind of test on it. We were happy. (It took so little to make us happy!) Then, starting on Monday, different folks kept coming up to us and saying, "You need a consultation," or, "We have to talk to you about hygiene."

"OK," we answered, "you can talk to us about hygiene on Thursday. We are scheduled for a lecture then." All this was a trifle unusual, and we commented that they were acting like we were scheduled for a test or something. To be on the safe side, we double-checked our written schedule to make sure that we were right about no test. There was none scheduled.

We arrived at the scheduled hygiene lecture on Thursday morning and immediately knew that it was not our ordinary lecture. The room was packed with all the hygiene-type instructors from Star City and a lot of other people we had never seen before. (Later, we determined they were from Energia, a Russian space company.) An elderly lady who appeared to be in charge started asking us questions. She wanted to know how we would determine the bacteria count on Mir. Seeing the Russian medical procedure book lying on the table, I said that I would use it to find a procedure to accomplish that task. I couldn't think of anything else to say because no one had ever talked to us about that subject. Then she started asking more questions along the same lines, and I gave the same answer, sometimes showing her in the index where I could find the procedure. An observer would have laughed.

I gathered that the elderly woman who asked the questions was hard of hearing, because each of my broken-Russian answers was screamed into her ear by a young female assistant. I had to bite my tongue to keep from laughing; the situation was just so funny. Finally, there was a question about what type of clothes would be provided for us in orbit. John replied that we had been asking that very question ever since arriving for our training. We had been expecting a session where we would be shown all the different possibilities of clothing and personal equipment, then choose what we wanted to use in orbit, just like we did for a shuttle flight. It never happened. After John reiterated that he really wanted to see what our choice of clothing in orbit would

Tumbleweed

be, a lady in the back of the room opened her purse and yanked out a set consisting of a skimpy white T-shirt and shorts. John asked when we would see the other items to choose from. The woman simply replied, "This is it." Once on Mir, I laughed as I thought back to this lady pulling out the shorts and shirt and waving them in the air. How truly she had spoken. We had seen all the clothing options!

Soon our scheduled two hours were finished, and we walked out of the chaotic session in the room. We turned to each other and said, "I think that we just had an exam!" and then really laughed. We both agreed that this was the best way to take Russian exams, with no warning, no time for test anxiety to build, and no long hours studying all those Russian vocabulary words!

For me, these tests were the major trauma of the training program. First of all, no one likes to look stupid. No matter how hard I studied or how well I understood the system in question, I would look dumb just because I could not express my knowledge coherently. I did not realize just how much this was getting to me until I woke up in a sweat one morning after a nightmare. In the dream, we were taking a test, and I was trying to answer in Russian. With each sentence I spoke, the agony on the faces of my examiners increased. Finally, one of them put his hands over his ears so that he would not have to listen to my broken Russian any longer. After I woke up, I realized that the good thing about the nightmare was that it was all in Russian. I had always heard that beginning to dream in a foreign language is a sign of finally becoming familiar with it!

The worst exam of all was on the plan of flight. This showed up on the schedule toward the end of the training flow, and I asked what it was. I was told not to worry about it, so I didn't. The testers were a few of the flight planners from the Russian mission control. The first question on this test was, "How many times is it planned for you to do the POSA experiment during your time on Mir?" I looked a little blank and replied that I didn't know. Then I was asked, "Where inside Mir will you do the DCAM experiment?" DCAM experiment? I had no idea what the acronym meant, so I answered again that I was sorry, but I didn't know. After about an hour of questions regarding the various NASA experiments and an hour of me not knowing any answers, the examiner said, "I can't believe you work for NASA and don't even know what experiments you'll be running on Mir." As I walked out of the test, totally frustrated, I muttered to myself, "I don't believe it

Lucid

either." How come the Russians knew what I'd be doing on Mir when I didn't even know?

Not being trained on the experiments I would be responsible for on Mir was to become a recurring headache of mine, one many people failed to understand. One evening, I had a call from a person from the Public Affairs Office (PAO) who worked at NASA headquarters in Washington, DC. She said that a reporter wanted to interview me about my upcoming flight on Mir. I replied that I'd be happy to do the interview but that I had no idea what I would be doing on the trip. I hadn't been able to get a list of the manifested payloads and answering a reporter's questions with "I don't know" might not make NASA look good. The PAO officer found it hard to believe that I had not been told what experiments I'd be conducting, much less been trained in them, and agreed to see if she could uncover any information for me. After hanging up, I pondered the mysteries of the universe—like how it was possible for PAO to always be able to find me and get a call through with no trouble when no one else at NASA, such as the people working the manifest, was able to contact me.

A week later, she called back and said, "Oh, Shannon, I have great news for you. I have found out what you will be doing!"

"Great, tell me, please!"

"Well, I checked with the chief scientist here at NASA headquarters, and he said that you would be the main engineer on Mir to reconfigure Priroda, the new module that will arrive while you are there, and to be in charge of all the Priroda systems."

I answered, "I don't know what I will be doing on Mir, but I know one thing for certain. My job is not to be the chief engineer for Priroda!"

At this point, you are probably thinking, like I was thinking, how it could possibly be that I was about to go on a spaceflight yet did not have a manifest of the science experiments that I was going to do. I can only report one small perspective on the story, and that is from the viewpoint of a crew member in Russia. When John and I arrived in Star City, it was just before the time of Norm's launch. Norm talked extensively with us about the problems and challenges of getting started in Star City. At that time, there were two or three full-time people stationed in the city to support the US science for Norm's flight. Of course, we talked with them a lot and listened to all of their "war stories" about the difficulties of training a US crew person in Star

Tumbleweed

City. As John and I listened, our eyes got bigger and bigger, and then we breathed a sigh of relief and said to each other, "Thank goodness we are not the first Americans over here training." We were so glad that Norm was the pioneer and that we, and the support folks for our missions, would be able to benefit from the experience of his flight. It was obvious to us that the learning curve had been so steep that it was a wonder anyone had been able to hang on!

Well, a funny thing happened on the way to Norm's launch. Now, you must remember, John and I spent all our days sitting in lectures and all our nights studying for the next day's lectures. We had no contact with anyone at NASA back in the States and, because of our schedules, very little contact with the few NASA folks living in Star City. So to us, it appeared that Norm launched, and then, without warning, the support folks were gone. No one was there working the science payload for our flights, it seemed. We eventually found out that contracts had changed and that the management structure of Phase 1 had been reorganized, but that was all learned later. All we saw was that there was no support and that, suddenly, instead of being recipients of the lessons learned from Norm's flight, we were reluctant pioneers—going through the training flow without the benefit of Norm's experience.

We knew that we could not influence the Russian training on the Russian systems. It would have been inappropriate for us to try. After all, we were there as paying guests. We were there to do the Russian training in the Russian way. But for the NASA science program, we naïvely expected that it would be conducted in a manner similar to that for a science mission on the shuttle. NASA management made the same assumption, which was part of the problem. The NASA way of doing things could not be transported intact over to Star City.

The training for the science we would be doing on Mir was done in several three-week blocks. I eagerly anticipated the first block of training, which was to be completed back in Houston. Not only would I be able to live at home for a few weeks, but I naïvely assumed that I would be doing training in English and in the familiar NASA way.

When I first saw the training schedule, I realized my assumptions had been faulty. The schedule said that training would begin at nine and end at six, with lunch at one o'clock. I protested that the typical Johnson Space Center schedule was to work from eight to five, with a break for lunch at eleven or eleven thirty. If my lunch hour was

Lucid

scheduled at one, I would not be able to eat with friends and coworkers, something I had been looking forward to during these three weeks of training in Houston. My protests fell on deaf ears. This was the way the Russians, who would also be doing the payload training, wanted it, so this was the way it would be done! Training always started at nine and ended at six, just like in Star City. Also, just like in Star City, I lunched alone.

The first time we had payload training at JSC was the first time I was to meet my Russian crewmates, Yuri Onufriyenko and Yuri Usachov. The Friday before training was to start on Monday, I asked the training folks to please introduce me to Yuri and Yuri as soon as we arrived on Monday morning.

On Monday morning at nine, I entered the room listed on the schedule for the first payload training session. It was full of American payload investigators, NASA and Russian training folks, translators, and cosmonauts. There were no introductions. I closely observed the knot of cosmonauts when they took a smoking break just outside the door. By the process of elimination, I figured out who I thought were Yuri and Yuri, and I went up to them and said, "Hi, my name is Shannon, and I think that you are Yuri and Yuri and that we are going to fly together."

They laughed and answered, "Yes, we are Yuri and Yuri, and we have also heard that we are going to be a crew together."

I then asked them, "Do you remember Ellen Baker?"

"Of course! No one could ever forget Ellen!"

"Ellen told me that both of you are great guys and that I am very lucky to get to fly with you."

They laughed. "We heard about you and Ellen and the 'Women of Atlantis.'"

Ellen and I had become good friends when we flew together on STS-34 and, together, deployed the Galileo satellite. Because we were the deployment team and because we were on the space shuttle *Atlantis*, our training team started calling us the "Women of Atlantis." The nickname stuck for a long time. Ellen had worked with Yuri and Yuri before because they were the backup for the Russian crew that Norm flew with and returned with on STS-71. Ellen was part of the STS-71 team. After she told me that Yuri and Yuri were great guys, I never again worried about crew compatibility during my Mir mission.

Tumbleweed

That morning, we started payload training. Translated, that meant that John, four cosmonauts, and I sat around a table and listened while the principal investigator (PI) for the experiment we were training on told us the theory behind the investigation. If we were fortunate, we also saw the equipment we would use. Although the PIs spoke English, each sentence was translated into Russian for the cosmonauts, so every lecture was double its normal length. During the translation, my mind would wander. By the time English was again being spoken, I would have forgotten what the last thought was. Training was painful!

After three weeks, it was back to Star City for more training—all without having time for lunch with my friends. But, on the bright side, I had spent evenings with my family.

Stage Two

The next block of science training took place in Star City and required that all the training hardware be shipped there. It was interesting to watch the accelerated aging process that occurred in the training folks whose responsibility it was to get the equipment through Russian customs and the Russian flight-certification process.

By the time this training session started, John and I had been able to convince the schedulers that the Americans and Russians needed to be scheduled separately for payload training. Having a large number of people gather around one piece of equipment, try to configure it properly, and go through the procedure did not do anyone much good. A person only learned if it was a hands-on session. Also, valuable time was wasted if you had to be there while it was explained in English and then translated into Russian. Time would be much better spent if the sessions were of shorter length and conducted in only one language. It was reluctantly agreed to let the Americans and the Russians have separate training sessions. Due to the logistics of getting all the equipment to Russia and also getting it from wherever it was kept in Star City to the training room, we were never able to do a procedure end to end, assembling the entire configuration as it would need to be done in orbit. However, the biggest problem of all was the procedures.

When we had our payload training sessions, we had the typical NASA procedures, and a procedure writer was always there to record suggestions on how to improve them. The procedures were in good shape, and I was happy. Then I found out that these great procedures were not the ones that we would use in orbit. This was beyond comprehension! The Russians would not allow NASA to use its standard procedure format for the NASA-built, NASA-funded project with NASA people working on it! These procedures, which were in a logical form, a form used in the NASA space program for years, could not be used as written. Instead, they had to be given to the Russians, who then not only translated them into Russian but also changed their format to the Russian version. Then these Russian procedures, in Russian format, were translated back into English, still retaining the Russian format, by the Russians.

The Russian format had much to commend it if you had to conserve paper and ensure that the procedure book did not weigh much, a valid concern for the Russians because everything they launched up to Mir had to arrive either on the Soyuz or a Progress vehicle. NASA had never had to operate under such rigid constraints, so its procedure format used more words and, hence, more paper, which resulted in more comprehensive and understandable procedures. In addition, many of the lines of Russian-formatted procedures read from right to left, not from left to right. This was the hardest thing for me to deal with. Whenever I started to use a procedure for the first time in orbit and came to a line that went from right to left, my aggravation index shot over the top. So I would just have to take a deep breath, count backward from ten, and start over. By trial and error, I found that the way that worked for me was to hook up all the various pieces of equipment in what I considered a logical manner and then use the procedure to check the hookup before applying power. This way, my aggravation index remained within reasonable bounds.

Undoubtedly, the biggest problem of all with the procedures was that I never saw a completed, final version of them before I launched. Every document I used in orbit was one that I was looking at for the first time! After I returned from Mir, I found out that the reason all this happened was due to the contract that NASA signed with the Russians for Phase 1, which had stipulated that the Russians would write all the procedures.

I will relate just one experience to illustrate the frustration of working with these procedures. John and I were scheduled for a training session on NASA hardware. Only a Russian instructor was present. The training object for that day was a piece of equipment that measured disturbance in the microgravity environment. The experiment itself was simple. All the crew member needed to do was place a sensor in the proper location called out by the ground, insert new optical disks, and turn the power on. Simple. I had conducted a similar experiment on one of my shuttle flights.

We arrived at the classroom and were given a procedure, Russian language only, of course. The box that contained the power switch and into which the new optical disks were placed was also there. We got to step two and stayed there for two hours because of a translation error. The person who had translated the text from English to Russian had made a small mistake with a preposition. The Russian procedure said to remove the old optical disk, then turn the switch *for* ninety minutes. What the English version correctly said was to turn the switch *to* the ninety (degree) position. We knew that holding a switch for ninety minutes was not the way the equipment worked, but trying to explain that to the Russian instructor, who spoke no English, with our limited Russian was almost impossible. In retrospect, the entire scene was quite comical, but the humor was not readily apparent at the time.

Interspersed among the mind-numbing days of eight solid hours of Russian lectures were a few practical sessions. One of the first was how to doff, don, and use the "Ckafander"—the launch-and-entry suit. Like in the shuttle program, the original flights of crews in the Soyuz were without any type of pressure suit. Because three members of a Russian team were killed due to the depressurization of the Soyuz during reentry, all crew members now wore a pressure suit for launch and entry. A similar evolution took place in the shuttle program, with crews wearing partial pressure suits after the *Challenger* accident. These suits protect the crew member in the event of a high-altitude bailout. We listened to a two-hour lecture about the Ckafander and were shown how to put it on, position ourselves in the Soyuz launch-and-entry seat, and strap ourselves in.

Suits and Seat Liners

Each cosmonaut and astronaut had a training suit for use in the Soyuz simulator sessions as well as a suit individually tailored to our own measurements to be used for flight. Norm used his suit because he had launched on the Soyuz, but the rest of the American Phase 1 crew members would use theirs only if there was a situation that would require an emergency deorbit in a Soyuz from Mir. Ours would be launched with us on the shuttle that brought us up to Mir. After being

Lucid

checked out in orbit, these suits would remain positioned in the Soyuz, ready for use in an emergency. Additionally, we each had an individual seat liner for our seats in the Soyuz. This liner was flown up to the Mir on the shuttle that brought us and returned, along with our Ckafanders, on the shuttle that brought us home. Each of these liners was individually molded to conform to our bodies.

After we learned how to doff and don the suits and strap ourselves into the seat, we were taken to the factory where the Soyuz spacecraft were manufactured, and we practiced getting into a real seat liner in a real vehicle. The main purpose of this exercise was to have us strap in and then measure how much clearance our knees had when the seat was in the landing position. (Just before the Soyuz contacts the ground on entry, small pyrotechnics fire and move the seat into a more upright position. The purpose of this is to cushion the crew member from the gravitational force experienced when the Soyuz contacts the ground.) When the seat moves, the crew members' knees are closer to the instrument panel. If the knees connect with the panel, the force of the contact could crush a crew member's bones. In a worst-case scenario, the situation might result in a severed artery, and the crew member could bleed to death within minutes.

After we strapped ourselves into a real Soyuz, the seat was pneumatically jolted into the landing position. Then, the clearance between knees and instrument panel was measured with a ruler. Fortunately, both John and I had enough clearance, although we were in the "gray" zone, meaning that there was not quite enough clearance for the engineers' satisfaction, but in all probability, our knees would not contact the instrument panel if we landed in a Soyuz. I thought it was interesting that although John and I are the same height, he had a little more clearance than I did. The deciding factor of whether a person could fit in the Soyuz was dependent not just on height but on femur length.

After it was determined that we would indeed fit into a Soyuz, we were taken to the factory where the seat liners and the Ckafanders were made by hand. We spent half a day in this facility, where, in previous years, some of the best of Soviet technology went into making ejection seats for Russian fighter aircraft. Now, to make ends meet, many of the engineers were designing and building skateboards.

We began the process by donning long white underwear, then sitting in a mold made of wet plaster of paris. It was sort of like an adult trying to fit into a baby bathtub. Just the main part of your body was in, and your head, arms, and legs all hung out. We would sit and then get out. After the technicians did a little scraping on the mold, we would sit again. This process was repeated until the technicians were satisfied. Then we were measured for our Ckafanders.

Months later in Star City, I passed Yuri Onufriyenko in the hall. He had heard that I would have a fit check for my seat liner and a fit check and pressure check for my pressure suit later that week. Informing me that the pressure could be uncomfortable, he advised me to make sure that all the "wrinkles" of the suit behind my knees were as straight as could be before it was pressurized. I said OK but wondered what he meant.

We returned to the factory one gray, cold, snowy day and were taken into the room where we donned the new pressure suits. These suits were literally handmade, and we met the woman who had spent many hours hand-stitching them. Then we sat in our individual seat liners. After the technicians checked to make sure that they were satisfied with the fit, cloth "caps" were placed over each knee, and the straps attached to these caps and also to the body of the liner were tightly cinched. As mentioned, this was to prevent our knees from crashing into the instrument panel upon the Soyuz's impact with the earth. I will just say it was uncomfortable and leave it at that.

Then came the pressure check. The visor was closed and locked. We did a comm check to make sure we could communicate with the suit technician, and then the suit was pressurized up to 4 psi. I found myself encased in a hard carapace. Only with difficulty could I flex my arms and legs just a tiny little bit. And then I understood Yuri's comment about the "wrinkles" behind my legs. After a few minutes, it felt like a dozen dull knives pressing into the backs of my knees. My legs went to sleep. My arms went to sleep. Unfortunately, my mind did not go to sleep. For two hours, I literally lay there trying to move just enough to relieve a little of the constricting pressure. It was such a relief when the test was over. To my dismay, I found out that it would be repeated all over again, only this time in an evacuated vacuum chamber. This only helped prove what John and I had already determined was the primary objective of the Russian training system: to teach the aspiring space explorer to endure pain. We joked that it was no accident that in English, the words *pain* and *train* rhyme! The Russian way—train equals pain!

Lucid

Although both the shuttle and Soyuz systems utilized pressure suits, the suits served different purposes. The shuttle suit was a partial pressure suit and the Soyuz suit was a full pressure suit, so it is not exactly fair to compare the two, but I liked the Soyuz suit because it was easier to put on and also because it was a more comfortable fit. Here again, this was because the Soyuz suit was custom-made for each individual, whereas the shuttle suit was not. As we were getting on the bus, ready to leave the factory after trying on our suits, an elderly Russian engineer pushed his way onto the bus and shook my hand and said that he was so happy and so relieved that my suit had fit. He said that he had not been able to sleep the night before because he had been worried it wouldn't. And then he said, "I was so worried because this was the first time that we ever had to make a suit for a large woman." I just laughed and told him to be happy. The suit fit great, and I was happy with it.

We had one more seat and suit-fit check before they were shipped to Florida to be loaded onto the shuttle, and that was in Baikonur, in the actual Soyuz that would bring Yuri and Yuri up to Mir. About two weeks before their actual launch, we all flew to Baikonur for a "practice" launch. This was the Russian equivalent of the terminal countdown test, or TCDT, that all shuttle crews participated in roughly three weeks before launch.

After supper on our first night in Baikonur, we went out to the hangar where the final preflight checks were being conducted on the Soyuz before being freighted by rail out to the launchpad. John and I went into the suit preparation room and put on our long underwear. I put on my Ckafander first, then went out to the hangar where the Soyuz was located while John put on his suit and the technicians checked it out. Waddling like a penguin, I climbed up flight after flight of stairs until I reached the level where there was a platform so that a person could go down through the top hatch. This was the hatch that would eventually open into Mir. Yuri Onufriyenko entered first, and then I slid down through the *bytovoi otcek* (BO), the living compartment, through another hatch, and into the *spuskaemy apparat* (SA), or landing vehicle, the crew compartment for launch and entry. I squeezed down into my individualized seat, sweat streaming from every pore. I couldn't remember ever having been this hot. As Yuri measured the clearance between my knee and the instrument panel, there was headshaking and discussion about the number of inches he totaled.

Tumbleweed

Gaylen Johnson, my NASA flight surgeon, decided that this would be a good time to lecture me on the gruesome consequences of not following the safety protocols exactly.

"Shannon, I want you to really understand how important it is for you to have your knees strapped down tight, so tight that you can't bear the pain, because if they are not, they will impact the instrument panel, and if that happens under the g loading of impact, they could be crushed, and not only that, if they are crushed, they could sever the artery located back there, and if that happens, there is nothing I can do for you! Do you really understand the importance of having those knees strapped down tight?"

I assured him that I did. Frankly, all I wanted at that moment was to get out of the Soyuz and then get out of the suit. At that point, I felt like I was locked up in a Russian sauna with no escape! The headshaking and discussions petered out, and I was told to exit. Finally, I was back in the suit room, peeling off the soggy suit and underwear.

As I sat in the suit room gulping down hot tea, I thought how glad I was that it was John's turn to be in the Soyuz now, not mine. Eventually, he came waddling back in and started to strip. Off came the Ckafander; then he warned me not to look because he was peeling off his long underwear. Because he ignored my suggestion that he step behind the privacy curtain, I did a 180 and faced the door.

So, there I was, sitting in a chair, drinking hot tea, facing the door. Behind me, I could hear John grunting and groaning as he was wiggling out of his sweaty suit and underwear. I could hear a low mummer of Russian as the various technicians were scurrying around behind me, helping John and working on both of our suits. Suddenly, General Glaskov, Deputy Chief of the Gagarin Cosmonaut Training Center, dressed in full military regalia, stood in the doorway. I saw his mouth open and drop a little and his face turn a deep red shade as he took in the scene. As he gazed beyond me to the naked American male, I wondered just what thoughts about "those crazy Americans" were careening through his head!

Water Survival

We had more time to become familiar with our suits during water survival. The theory here was that because the Soyuz was our lifeboat on Mir, we might have to be ready to depart at a moment's notice—and not in a nice, orderly fashion that would ensure a touchdown in

Lucid

the vicinity of Baikonur. We might end up in an ocean, a lake, or the frozen tundra, so we were trained to survive in any of these situations until we could be rescued.

We did water survival during the last part of May and, coincidentally, just when the Moscow area was experiencing a heat wave—note that "heat wave" in the Moscow area meant the temperatures got up to the high eighties, not over one hundred for weeks on end, the meaning of "heat wave" in Oklahoma.

Several different crews that were in training—instructors; technicians; doctors, American and Russian; and various family members of the Russian instructors and technicians—all got on buses one morning and headed out to a small lake about an hour's drive from Star City. A small village was located nearby, and as soon as the buses arrived, many of the villagers started wandering up along various paths through the forest to join our group. The entire scene was permeated with the festive air of a holiday picnic—sort of a Fourth of July celebration without the fireworks.

In the small lake, which was a reddish color, presumably due to some type of algae growing there, bobbed a life-size replica of the Soyuz—the craft the Russians use for launch and entry. It is similar in size and shape to the craft used by American astronauts when they launched to the moon. It was tied to the end of a rickety wooden pier. We exercised great care in walking out to the model for our initial briefing. Our launch-and-entry suits were unloaded from the bus as we stood around in the heat, waving the small branches that we had torn off bushes to continually swish over us to get rid of the gigantic horseflies that were having a feast on our exposed body parts. A tent was hastily erected in which we could change our clothes, and we put on our long underwear and the launch-and-entry suits.

After changing, John, Vasily Tsibliyev,-an experienced Russian cosmonaut who was our commander for the exercise-and I climbed into the capsule and strapped ourselves into the seats. The hatch was closed. Immediately, I thought that I would die from the heat. And it only got hotter! While the instructors on the outside were rocking the capsule to make it a realistic representation of being tossed to and fro by waves in the ocean, our task was to get out of our launch-and-entry suits and dress in the gear meant for winter survival in the Arctic Ocean. This consisted of a woolen sweater, a woolen coverall, a woolen jacket, a woolen hat, and a down-filled coverall. After all that gear was on, we had to put on a rubberized suit that would keep us dry. Then, we were to open the hatch and jump out of the capsule and into the lake. Easier said than done!

Picture three people locked inside a Volkswagen Bug, windows rolled up tight, parked in the parking lot of the Johnson Space Center in Houston, Texas, in July, who are then told to change their clothes and put on Arctic survival gear. That is the exercise we were doing.

Three people inside the Soyuz is a very tight fit. It is the type of fit where one person does not move, even a little, without thinking about his or her neighbor. My first thought after we closed the hatch was, "And we are supposed to change our clothes in here?" I could have sworn that there was no way to accomplish the task. Luckily, Vasily had done it before and was an accomplished master of the feat. He had us, one by one, doff and don various segments of clothing. While one person was getting his or her arms out of the suit, the other two were leaning in the opposite direction so there would be room for that person to maneuver. We kept doing this bit by bit until we had our suits off. We followed the same process in reverse to put on the various layers of clothing, all the while getting hotter and hotter. We were totally focused on trying to finish the exercise. We had no excess energy to wave at the village kids who had swum up and, noses pressed flat, were peering in at us through the small Soyuz porthole.

To make matters worse, John was an uncommonly heavy sweater. He is the type of person who looks like he has performed an Olympic feat, from a sweatiness standpoint, after just standing up. (He told me that he had often used this to his advantage in the beginning of his military career when doing mandatory exercises. The instructor, judging solely by the quantity of sweat produced, regarded him as an exemplary achiever and held him up as someone the rest of the platoon should emulate!) But in the Soyuz in the middle of a Russian heat wave, the sweat was a problem for Vasily and me. The Soyuz was tilted in the water. My seat was the lowest of the three. John's was on the high side, and Vasily was in the middle. John's sweat poured over us like a waterfall. It was hard enough trying to get out of the launch-and-entry suit when it was dry, and it was virtually impossible when it and any of our exposed skin was sopping wet.

We finally made it, throwing open the hatch and then rapidly jumping one after another into the lake. The coolness of the water could be sensed through all the layers of survival gear, and it just felt so good!

Then it was into the capsule one more time, only this time, we simply had to fasten on the flotation gear and jump out into the water while still in the launch-and-entry suit. Walking out of the lake with water pouring off me and thinking that I was in the beginning stages of heat exhaustion, I knew I never wanted to be that hot again.

Stripped down to our soaking-wet white long underwear, we all sat around and ate our lunch of boiled potatoes and boiled beef, mindlessly swishing our branches against the horseflies and letting the holiday buzz of the villagers seep into our subconscious.

Later, we again put on the launch-and-entry suits and climbed into the capsule, but only for a few baking moments this time. Then, we jumped out and slithered into a rubber raft. After practicing the use of flares and getting all our rafts floating in a nice line together, we jumped into the water once again and floated there, waiting for a helicopter pickup. I was near the end of the row of drifting bodies to be picked up, so I had plenty of time to gaze around. As a gigantic purple thunderstorm approached, I wished I knew the weather conditions acceptable for Russian training! Watching the lightning and thinking that it was certainly close enough to call off any type of aviation or water exercise, I struggled to keep my head above water as the hovering helicopter approached for my pickup. Finally, I had the clip from the helicopter line hooked into the ring on my chest and gave a thumbs-up, signifying that I was ready to be hoisted up and into the helicopter. Something got stuck midway through my upward journey, and I just hung there, spread-eagle, against the lightning-illuminated black sky. A picture from my fifth-grade history book of Benjamin Franklin hanging on to his kite as it sailed in the stormy sky, electricity sparking around his hand, popped into my head. Not wanting to think too much about the possibility of becoming another famous electricity experiment, I turned my head and gazed down at the golden onion-top dome of the small Russian Orthodox church glittering under the brilliant blue sky and dazzling sun, located way beneath me and several miles to the right. Never in my life had I imagined that one day I would be dangling from a Russian helicopter over the Russian countryside, transfixed by the beauty of a Russian church. Almost too soon, I started being winched up again and was quickly pulled into the helicopter.

Tumbleweed

Becoming a Crew

From the beginning of our arrival in Star City, John and I had been looking forward to the time when we would start to train with "our" crews and not just with each other. Of course, our concept of integrated crew training was based on our previous experience with shuttle crews. For the year before a shuttle flight, your crew, in many ways, became your work family. By doing flight simulations together day in and day out, going to meetings together, and traveling to various parts of the country to work with the different payloads that were on the flight, we came to know the members of our crews very well. We also learned to work together as a team, as a unit. This opportunity to build an assortment of assigned people into an integrated unit, a team, and a crew is one of the most satisfying aspects of spaceflight.

Finally, the week in late October arrived when we received a schedule showing that we would be in a Soyuz simulation with our respective crews the following week.

I was excited. I would get out of the classroom lectures and into real space-training stuff. Very rapidly, I realized that this crew training was not done in the NASA style and was not what I had expected.

The first time that I entered the area of the building containing the Soyuz trainer, I was struck by how quiet and empty the place was. In this huge hangar-like room, there were only two Soyuz mock-ups. Only one was used routinely for launch-and-entry training. We used the other one only once for a communication training session. Along one wall of this giant room were several smaller rooms. One was where we changed from our street clothes into our Ckafanders. The person in charge of controlling the computers that were running the training profile in the Soyuz sat in another of the rooms. Generally, I rarely saw more than three people in this entire area every time we were there for a training session.

High along the wall in this room, much higher than eye level, were photographs of all the Russian and guest cosmonauts who had flown in space on Russian hardware, along with their names, the flight name, and the flight duration. Each Russian commander picked a name, maybe that of a star, a bird, or something that had significance for him, and this was the call sign of the crew. Yuri chose "Skiffer" to be used as the call sign of our Mir 21 flight. This was the name of an indigenous people who lived in the southern forests long ago in Russia. When I

Lucid

asked further questions to find out more about them, Yuri replied, "They are very much like the Indians in America, Shannon." The idea that, soon, my picture would be included in that august stretch of photos was almost beyond my comprehension.

In the Soyuz, I was only doing entry training with my crew; this training was for the possibility of an emergency situation aboard Mir that would force us to do an emergency evacuation and emergency deorbit. I did not do any Soyuz training for launch because I would be launching on the shuttle. I had been trained to use the comm panels and discovered that I could be of some use because I had long arms and could reach the onboard switches more easily than Yuri could. When he pointed to a switch, I completed the required action. Despite the usefulness of my long arms, I rapidly discovered that my main job during this training was to scrunch up on my side of the Soyuz and be quiet, being sure not to get in Yuri's or Yuri's way or distract them from getting us safely home.

After the first few training sessions, I began to feel more confident, not because I fully understood what was going on but because I sensed that Yuri and Yuri were always on top of every situation. Also, they were always kind to me. Never once in our training sessions together did they make me feel like I was excess baggage. Yuri Onufriyenko, the commander, always included me in the cockpit activities so that I felt like I was part of the crew.

The simulator sessions were different from shuttle simulator sessions, and it was not just the difference in vehicles. For example, the "living section" of the vehicle, which separates from the crew compartment early in the reentry profile, did not separate during one simulation. This forced us into a different set of procedures. After the sim, I asked the instructor why the living section had not separated. He looked at me blankly. I repeated the question, trying to say that I wanted to know why the two did not separate. In my mind, I was thinking, "Can this happen because of a software failure, a hardware failure, or what?"

The instructor looked at me for a few more seconds and then just simply said, "Because that is the way that I did it."

I never did discover a "real world" situation that would have this result. The problems that come up in shuttle simulations are always linked to events that might really happen, and the probable causes for these situations are typically explained.

Tumbleweed

In addition to the Soyuz sims, we spent two half days in the mockup of the Mir Base Block and also completed a daylong exercise in the mock-up of part of the Base Block that was located inside a vacuum chamber. At the end of the day, we went through the procedures that we would do on Mir if there were a depressurization and we needed to get into the Soyuz for an emergency return to Earth. This was the sum total of the simulations we did on orbit training.

After we finished a Soyuz sim one day, we were scheduled to have tea with the psychology group. I'd had several meetings already with the psychologists. Previously, they had me do a few exercises, such as draw a picture of my home. Not being much of an artist, I drew a stick house surrounded by stick flowers and some stick figures, with one on the roof. When asked about the meaning of having someone from my family on the roof, I replied, "Oh, that is my son—when I am home, I can usually find him on the roof of the house or some other unlikely place." Another time I was given a stack of various cardboard chips with nothing but a solid color on one side and told to arrange them in my order of preference. I quickly removed all the shades of brown, gray, and black, saying that I did not like them at all and would not arrange them in any order because they were absolutely unsuitable. The psychologists told me that I needed to have them in order, but I insisted on leaving them out. I never heard what their conclusions were from these types of tests. No one ever complained, so I figured that they would fly anyone as long as the Americans were paying!

One day Yuri, Yuri, and I walked over to the "psychology session" together in the late afternoon, already dark because of a sky full of gray clouds heavily laden with snow. While watching every step so as not to slip on the ice, Yuri Onufriyenko said to me, "Shannon, we are going to be with the psychologists now. If they want us to do something, just remember to work together with us and agree with everything that they say. You understand?"

I laughed and answered that I did. To myself, I said, "You bet I understand. What I understand is that pilots the world over share a common culture, and this culture transcends their nationalistic culture!"

Soon we were all cozily sitting in an informal circle in the psychologists' office. We had tea. We talked pleasantries—or rather, I should say, the psychologists talked, and the Skiffer crew muttered appropriate nondescript responses to keep the conversation going.

Lucid

Then one of the psychologists handed each of us a box with a dial containing a semicircle of numbers and a needle that could be controlled by a black knob. Each of our boxes was connected to a larger black box. He then said that we were to work together and get all of our needles centered at the same time. He flipped a switch on the main black box. I rotated my knob, and the needle stayed in the middle. After trying this several more times, I said, "I think this experiment is broken, because no knob movements have any effect on the movement of the needle." Immediately, all our boxes were taken up and whisked away with the main black box. We drank more tea and then some more tea, and finally we left and walked out into the black night filled with blowing snowflakes. We never had another meeting with the psychologists, and I never did find out if the black box and the dials were broken or not.

What I found the hardest about the training regimen in Star City was the complete lack of control over the daily schedule and the total lack of time during the working day to take care of necessary personal matters, such as exchanging money, making travel arrangements, or even such a simple thing as calling to schedule repairs to be done to my apartment. When training for a shuttle flight, the crew received a preliminary schedule the week before so that they could make their inputs to the schedule before it was finalized. The crew members were asked about what training they thought should be emphasized and whether certain topics should be repeated. The entire training flow was a heavily crew-interactive process. It was not this way in Star City at all. Never once were we asked for any inputs to our training schedule. Never once were we allowed to take a few hours of vacation during a working day to do something of a personal nature.

I asked for only two hours off the entire time I was training in Star City, and it was not allowed. My daughter was coming to Russia for the first time and was nervous about arriving at the Moscow airport without someone being there to meet her. I asked everyone I knew if I could have the last two hours of her arrival day off so I could go to the airport to meet her. I thought it was all settled until the schedule showed up for that week, and it showed we were scheduled for food tasting at exactly the time I had asked to have off.

If it had been any other scheduled subject, I would have just not shown up. But I had to go to the food-tasting session. Food had been a big political topic between NASA and the Russians. NASA had been

Tumbleweed

insisting on the crews having the opportunity to taste planned foods before they flew. The Russians had been resisting the idea ever since it was suggested, and they had finally agreed. I always felt that they scheduled it at the time they knew I was trying to get off just so that if I didn't show up, they could say to NASA, "See, we tried to provide an opportunity for your crew members to taste the food, but they refused to participate!" Not being able to face the political fallout from not showing up for the food-tasting opportunity, I told my daughter that she would need to have an "arriving in Russia adventure" on her own. Fortunately, Brenda Blaha came to her rescue, because at the last moment, it turned out that she was having friends arrive at almost the same time as my daughter, so she kindly volunteered to go early and meet her at the airport.

August was our favorite training month—not because of what we were studying but what we were not studying! August is vacation month in Russia, and at times, Star City seemed almost like a ghost town. But when we had asked about going home in August, the answer was, "No—there is too much training to do!" So there we were doing training—or not doing it, depending on your definition. For instance, we would show up for training in venipuncture—learning how to draw blood—and look at the needles for about five minutes. No subjects could be found, so training went no further. This became our routine throughout August. We never complained—after all, it gave us a few more hours to try to get caught up on our notes! We went through the Soyuz comm classes several times. Again, it was fine with us; it gave us more time to try to catch up.

When autumn came, there were a few days of such incredibly colored intensity that I thought I was strolling to class on heavenly streets of gold. But all too soon, the seasons cycled back to winter. The skies were gray, and all paths were covered with snow. Days were short. We had lived in Star City for a year. We had the finish line—launch—in sight. But then it seemed to John and me that our world caved in.

We had thought that we were coming along fairly well with our training. We sort of had a handle on what was going to happen and what to expect. All of a sudden, our instructors started reeling off lists of subjects we were going to have lectures on and, what really bothered us, lists of more oral tests. Didn't anyone like us anymore? We had

Lucid

been going back and forth to class under the impression that our instructors liked us and that all was going along OK.

One day, I found myself in a class with an instructor, and he was telling me the names of bolts that were on the launch rocket that held a covering in place. I asked a simple question: "Why? Why are you telling me this?"

His answer was, "Because you need to know the names for your test!"

"Test!" I exclaimed. "What test? Why am I listening to lectures on the launch capabilities of the Soyuz? I am not going to launch on the Soyuz! I am going to launch on the shuttle!"

"That is not important," he replied. "You have to know the names of these bolts!"

Arguing was getting me nowhere. For a minute there, I had forgotten that I was in Russia and had reverted to my US classroom style of asking questions, arguing points, and expecting rational replies! Walking back under the gray and gloomy end-of-the-day sky, I met up with John and we compared notes. He was experiencing the same type of irrationality. I thought of my return ticket tucked away safely in my sock drawer and received comfort from thinking that only a phone call to reserve a seat was between me and home. But before we exercised that option, we went and talked to Charlie Precourt, the most recent astronaut to be rotated into the position of director of operations for NASA in Star City. We told him what was happening and begged him to find out the reason. And he did.

It just so happened that NASA and the Russians had been having high-level meetings—meetings that involved scores of personnel from both sides. One topic on the agenda was a discussion of Norm's debrief items, one of which was that because the American crew member would live on Mir for months and not just for the usual one- or two-week guest cosmonaut visit, he or she should be given more Mir-related duties. This was a very reasonable point. The Russians said that from then on, the American crew member would have the official title of "board engineer two." (The Russian crew members were the commander and board engineer one.) The Americans interpreted this as meaning that the NASA crew person would have more duties on Mir. The Russians interpreted this as meaning "more tests for the Americans." Charlie figured out this discrepancy in interpretations. He was finally able to negotiate with the Russians and obtain a specific list

Tumbleweed

of topics that we were to be tested on in order to be called "board engineer two," and he secured an agreement that no more tests would be randomly thrown at us. John and I were very grateful to Charlie and totally mystified by how NASA could charge into situations without consulting the people who would be affected by policy changes—in our case, the crew in training, John and me. This was a case where good intentions backfired exponentially!

During training, the total lack of human engineering in the construction of Russian spacecraft added extra complexity to an already complex task. I had flown airplanes for thirty years by the time I got to Russia, and over those years, I had just accepted certain things in aircraft design as God-given rights. For instance, I expected size and shape to tell me something about the importance of a knob or switch. I expected the color of knobs and switches to tell me something, like red is important or has something to do with safety. I expected safety equipment, such as fire extinguishers, to be highlighted in red, to stand out in stark contrast to the environment. Aircraft panels are generally laid out in a similar manner. Going from left to right, I can determine the information needed when I get into a new plane for the first time, and I expected a similar degree of consistency in a spacecraft.

None of these assumptions worked in the Mir and Soyuz trainers. The oxygen masks and the fire extinguishers were a dullish gray that blended in perfectly with the gray-blue panels. The panels were great camouflage for them! Some switches and knobs that were of utmost importance in critical situations were very small and in barely accessible places. One afternoon, I had a Soyuz familiarization course. During the discussion of the function of various switches, I asked what one tiny switch I could barely see was there for. I was told the acronym for the switch, and I then asked to have it repeated because the acronym was identical to that for another small switch I was already familiar with. Then I asked point-blank if the switches had the same Cyrillic letters for their names because I thought I was misunderstanding something. I had not misunderstood. The switches were labeled the same. Both were tiny, and both were of the utmost importance in very different emergency situations. One was used if there was a depressurization, and the other was used if a hard landing was anticipated. Two completely different emergencies. Both were extremely hard to reach. The board engineer one would have to reach

Lucid

around his or her leg to get to the appropriate switch and feel for it with his or her gloved hand. Unbelievable!

STS-76 Crew Arrives

John and I had eagerly anticipated the arrival of the STS-76 crew, the crew with whom I would launch to Mir. Not only would it be great to see our friends, but we figured it would be a break from the daily routine of lectures. (We also thought that we would be able to participate in some training where English would be spoken because the shuttle crew was not required to know Russian.) We were wrong. The STS-76 crew was treated like a visiting American crew—read, *royalty*—while John and I still trudged, by ourselves, to our typical lectures. We jealously watched the STS-76 team being transported back and forth in a van to various training functions that we had no part in. The low point came when the STS-76 and Mir crews were being introduced to Star City management and I was sitting in the back of the room with my Russian teacher watching the proceedings, not being included in either crew. Neither the Russians nor the Americans knew what to do with the American/Mir crew members. We were truly neither fish nor fowl!

Final Exam

The launch date for Yuri and Yuri was rapidly approaching, which meant it was time for the final exam that would certify our crew to fly. The testing would occur in both the Soyuz and Mir simulators. Even though I had no doubts about my ability to perform well as a crew member in either the Soyuz, if necessary, or on Mir, I felt a lot of pressure. I really did not understand what was going to take place during these exams. My personal goal was to make sure that I didn't make Yuri and Yuri look bad.

The morning for the Soyuz test arrived, and I went through my usual morning routine, leaving the apartment a few minutes earlier than usual so that I wouldn't make Yuri and Yuri nervous by having to wait for me. As I walked into the simulator building, Gaylen practically ran me over as he came out of the glass doors.

"Where have you been?" he asked. "I've been sent to look for you."

"Look for me?" I answered in surprise. "I'm at least ten minutes early!"

"No, you aren't," he replied. "It's already eight forty!"

Tumbleweed

Well, I was right in a sense. I was early for nine, but not eight. For some reason, this test had been scheduled to begin at eight. I hadn't even looked at the time we were to start. I had just assumed that it would be at nine like always!

Gaylen helped me put on the launch-and-entry suit, and then I rushed out to the table in front of the Soyuz trainer. Yuri and Yuri were already there, all suited up. I mentally told them I was sorry for making them wait for me while they stood there getting hotter and hotter in their suits. The Russians in charge of our training were also waiting, all dressed in their finest blue Russian Air Force military uniforms. On the table were several sealed envelopes. After a few introductory words from the ranking officer, he instructed the crew to select one of the sealed envelopes he held out. Each one contained a separate scenario. Yuri Onufriyenko asked me to pick. Cringing slightly with the responsibility of choosing the test case, I reluctantly reached out to choose an envelope. I touched one but then took the one next to it. After I handed it to the officer, he passed it on to the simulator instructor. Yuri, Yuri, and I crawled into the Soyuz to await whatever catastrophes would come from that small envelope.

The simulation went well. Every malfunction was handled promptly by Yuri and Yuri with graceful ease. I was proud to be part of the Skiffer crew.

Once the simulation ended, we put on our street clothes before meeting with all the high officials and instructors in an adjoining room. There, every action and response made during the sim was dissected and analyzed before the officials admitted, grudgingly, that the crew had done an outstanding job. I was struck once again by cultural differences. At the end of a shuttle simulation, there were, of course, extensive debriefs. These were aimed at making sure everyone understood what had occurred and what could be done better next time. Generally, it was a positive, helpful atmosphere. At this Soyuz debrief, positive comments seemed to be against protocol. It was as though no one wanted to say anything good about anybody. Eventually, though, the officer in charge said the magic words: the Skiffer crew was ready to fly.

I left the building alone and gazed up into the cold, black sky, brilliantly illuminated by a multitude of stars. I was now a certified board two engineer. The commission had said that I was ready to launch on the shuttle and live on Mir). From now on, it would be just

Lucid

a matter of waiting. That evening, after I had finally slid my way back to my apartment, I did not do any useful work. I just sat around, gazed out the window at the sky, and kept repeating, "You made it. You are certified to fly. No more tests." I sort of expected the phone to ring with someone from NASA calling to say, "Hey, we heard you made it! Congratulations!" The phone didn't ring.

LAUNCH

Norm Thagard, the first American to live on Mir, launched with his Russian crewmates Vladimir Dezhurov and Gennady Strekalov from Baikonur, Kazakhstan. The shuttle brought him and his crewmates back to Kennedy Space Center (KSC). Norm was the only American who arrived at, lived on, and returned from Mir with the same crew. All the succeeding six American astronauts to live on Mir left Earth and returned to Earth on the shuttle, whereas their Mir Russian crewmates arrived on Mir and returned to Baikonur via a Soyuz vehicle. With the completion of the Mir 21 training, it was time for Yuri, Yuri, and me to part ways for a month in order to be reunited on Mir by means of separate spacecraft.

The first step in the journey of Yuri and Yuri to Mir occurred early in the morning a few days before the February 21 launch date. They, the backup crews, the surgeons, the suit technicians, many more people whose exact function I never quite figured out, and John and I boarded several large planes at the Chkalovsky airport for the flight to Baikonur. Once we all settled into our seats, there was suddenly, as if in response to some telepathic signal, a mad scramble for the doors as all the males deplaned. I, the lone female, was the only one left on board. I looked out the window to figure out what was going on, and under the brilliant blue canopy of the winter sky, I saw a line of male backs and more than fifty brilliantly golden arcs descending into the

Lucid

snowbanks that lined the runway. Then I understood the hasty exit. Everyone had realized that there were no toilet facilities on the plane. But I was not worried. I had lived in Russia long enough to know not to drink much in the morning if I did not for sure know where the next restroom would be!

Before the launch, we stayed in the Russian crew quarters, a hybrid dormitory-hotel building isolated by an intimidating fence on the outskirts of the town of Baikonur, known as Leninsk until 1995. Behind it was a grove of trees, each placarded with the name of a cosmonaut. We found the names of the Apollo-Soyuz crews, and then it seemed that no more trees had been planted. When we asked why, we were told that Russia no longer had the money to plant memorial trees. There would be no memorial trees for Yuri and Yuri and certainly not for John and me!

The night before the launch, the crew and assorted high-ranking officials gathered in a large room. The prime crew, backup crew, and John and I sat behind a glass partition on an elevated stage. My guess was that the glass was a form of quarantine, never mind the fact that we had been mixing with the people on the other side of the glass just minutes before. Everyone in the room could watch us. Then the lights dimmed, and the movie *White Sun of the Desert* started to roll. We watched. This was a long-standing Russian tradition the night before launch. Superstition held that disaster awaited the launch of the crew that did not watch this movie. Vladimir Titov's crew, the only one to not watch this movie since its filming, had to eject thirty seconds after liftoff.

I spent the night in the room that Norm had slept in the night before his launch on the Soyuz. Every time I walked out the door, I looked at his name, which he had, in the tradition of cosmonauts leaving their rooms on launch morning, scribbled on his door with a black Magic Marker.

Launch morning, the buses headed for the launchpad were loaded in front of the crew quarters. On the first bus was the prime crew. I was considered a member of this crew for a short while longer. The various support personnel, the flight surgeon, the suit technicians, and others also rode on this bus. The buses drove slowly across the bleak, cold steppes of Kazakhstan. As we rode along the deeply potholed blacktop road, TV monitors in the front of the bus played a special tape for Yuri and Yuri. It featured their friends and families saying a

Tumbleweed

mournful goodbye. Yuri Onufriyenko's wife sang a plaintive song, accompanied by his teenage son on guitar. Furtively, I dabbed the corners of my eyes with a wadded-up Kleenex. Inside the bus, there was no jesting, no small talk, no smiling. We drove at funeral-procession pace through a gray village where the gray inhabitants solemnly lined up in their gray doorways in their gray clothes.

We finally arrived at the building in which the cosmonauts were to be suited in their full pressure suits. These suits would protect them if, for some reason, the Soyuz depressurized after launch. The press and all the rest of us kibitzers were ushered into a room with seats facing a wall dominated by a gigantic picture window covered by a curtain. We crowded into the room and jostled into the seats. Then, we waited. Hushed whispers about irrelevant details were exchanged with our neighbors. It was as though we were in church, waiting for the start of High Mass. We continued to wait. Finally, the curtains opened, and we saw Yuri and Yuri sitting in chairs, dressed in their pressure suits, with their backup crew members standing behind them. No one smiled. One by one, various members of the audience, including myself, went up to the glass partition, sat in the provided chair, said a few words of greeting, and wished the crew Godspeed.

Shortly thereafter, it was time for the final segment of the launch procession, so the crowd all filed out and waited outside in the subzero temperature for the crew to appear. When the crew members exited the building, they walked to the squares that had been painted on the cement long ago, stood at attention, saluted, and received their commission to go and launch. Then it was back into the bus and onward to the launchpad. Just before arriving at the pad, there was a quick stop for the ritual urination on the bus wheel. Yuri Gagarin had done this on his way out to the pad for the very first Soviet launch, and every crew since then has followed this tradition.

The buses stopped on the pad, just a few hundred yards from the fueled rocket. I watched Yuri and Yuri, clad in their blue-trimmed white space suits, climb off the bus and waddle to the scaffolding. After they had climbed a few steps, they turned and waved farewell to all of us, then entered the elevator and ascended to the hatch to mount their cosmic chariot.

The buses emptied quickly as everyone jumped out to snap a few photographs. One of the backup cosmonauts lit a cigarette, seemingly oblivious to the fact that a fueled rocket stack stood just hundreds of

Lucid

yards away. A pad worker quickly knocked it from the cosmonaut's lips and ground it into the cement. Then we all piled back into the bus and headed for the bunkers from which we would watch the launch.

John and I waited together in the viewing area, stomping our feet to keep our circulation going in the arctic temperatures. In vain, we looked around for a place to stand where we would be sheltered from the wind. Luckily, the backup crew spotted us and graciously invited us into a bunker. After a snack of small sandwiches and hot tea, we walked back out to continue waiting and enduring the cold. The rocket looked like a single skinny pencil propped up against the sky. A tattered banner extolling the glories of the Soviet worker flapped against a fence that, although originally intended for crowd control, would now have been no challenge to any self-respecting cow. The countdown reached zero, and the skinny pencil on the horizon lifted upward on a triangle of orange flame. There was no thunderous shaking of the earth, no gigantic, billowing, smoky white footprints lingering in the sky, as there was at a shuttle launch.

John and I returned to the bus. We retraced our route back to Baikonur, but this time it was a party atmosphere. Champagne, fresh fruit, and bread were passed around the bus. We bumped over the asphalt road across the Asiatic steppes, celebrating the successful ascent of our friends.

Arriving back at the hotel, we learned we could not return immediately to Star City that night because the entire Moscow region was gripped in a vicious ice storm. Instead, we ate supper at the table where Yuri and Yuri had sat just hours ago, but this time, the seats were left empty in their honor. Halfway through the meal, a NASA person came in and unwittingly sat in Yuri Onufriyenko's empty seat and ate. Although this breach of tradition obviously upset the Russians in the dining area, no one said anything.

The next day, when John and I finally made it back to the Chkalovsky airport, I exited the plane onto the ice-covered stairs. Immediately, both feet flew out from under me, and I landed on the top step in a sitting position. Very carefully, I inched into an upright position and then almost crawled down the steps to the ground. A month from launch was not my idea of an opportune time to break a leg! Besides that, we still had a few odds and ends of training that had to be completed.

Tumbleweed

I had been pushing very hard to be allowed to leave Russia right after the Yuris' launch and return to Texas so that I would have the last weekend in February at home. I did not think it would be unreasonable to have one weekend with my family before leaving them behind on Earth for an anticipated four months. The Russians thought otherwise and insisted I stay in Russia that weekend and leave the following Tuesday. There was "very important" training to be done on that Monday, and it was imperative that I not miss it. I stayed.

On Monday, I went to a lecture and demonstration of a computer in someone's office. This computer had nothing to do with spaceflight; it was not even one of the computers on Mir. Then, John and I attended a lecture on how we were not to eat any of the American canned fruit that would be on Mir. The Russians had not approved these cans. I simply listened in amazement, trying not to become angry that I had missed precious hours at home because of this "vital" training. Finally, six o'clock came, and training was really finished. I rushed to my apartment and finished packing just before the NASA van arrived at four in the morning to bring me to the airport.

Twenty hours later, groggy with sleep, I deplaned at Hobby Airport in Houston. I was thrilled to find not only my family but the entire STS-76 crew, plus the children of Kevin Chilton, who was the commander of STS-76, and several of their friends. The children playfully fought over who had the privilege of holding my hands as we wandered to the baggage area. It was good to be home.

The next morning, I was swept into the mad swirl of closely choreographed activities that consume the weeks preceding a shuttle launch. In between flying to KSC for the terminal countdown test (which is a dress rehearsal for launch), the last simulator sessions, and the myriad of last-minute very important meetings, I had to squeeze in a day trip to Oklahoma City to say goodbye to my mother and father. After a week of quarantine at crew quarters located at the Johnson Space Center, the six of us climbed into T-38s and flew in formation down to KSC in anticipation of launching on March 21.

In January, Michael had asked me if it would be all right if he missed the launch. One of his college classes was going on a field trip to Costa Rica. Unfortunately, they would land back in Florida on March 21, a date that had held steady as the launch date of STS-76 for over a year. I told him it would be fine. He said he probably would not miss the launch anyway because I had never been on a flight that launched on

Lucid

time. I laughed at his bold assurance. When I said goodbye to him as he left for Costa Rica, he was just a little puzzled that the launch date had not yet slipped.

His calculated gamble paid off. The evening of March 20, it was decided to slip the launch one day due to the forecasted high winds at KSC. When Michael landed in Miami on March 21, the first thing he did was call the number I had given him to determine the status of the launch. When he was told that it had been slipped one day, he replied, "All right! I knew NASA would not fail me!"

The person on the other end of the conversation was slightly taken aback at this unorthodox reaction to the news of a launch delay and replied, "Well, sir, we do aim to please!"

The last night in crew quarters just before a launch is somewhat surreal. I have always felt suspended between two worlds, one foot planted firmly on the earth and concentrating on the goodbyes, the other foot floating free, already stepping toward the approaching adventure! The night before a launch, I always called my parents and family to say good night and goodbye. That last call with my parents inevitably turned nostalgic. My father's voice reminded me of the first time I took him flying. He looked at me as I was proudly showing off my aeronautical knowledge and, in wonderment, said, "You know, Shannon, when I first laid eyes on you right after you were born, I never dreamed that this is the way you would turn out."

That night when I called my daughters, we never even got around to the goodbyes. We spent the time laughing about the poem they had composed while flying to Orlando the previous day. In my mind's eye, I saw them, one in the window seat, one in the aisle. They would be leaning over Jeff, Shani's husband, who consistently flew in the middle seat between them, both girls busily batting his personal pillows, which he always traveled with, out of the way so that they could see each other. As I pictured them giggling together, creating their limerick, I found it hard to remember that they were grown and no longer the teenagers I had parented only yesterday. My husband had hand-delivered the poem to me at supper, and I assured them it would be tucked in the biomedical pocket of my launch-and-entry suit when I traveled out to the pad in a few hours. I hung up the phone and laughingly reread their creative masterpiece:

The Astronaut Mother

There once was an astronaut
mother,
Whom Earth had started to smother.
So, she went up on Mir,
Without any fear,
Gladly forsaking all others.

But once she arrived up on Mir,
She yelled, "Somebody's already here!"
Then the two Yuris popped out
And said with a shout,
"We're living with you for a year!"

The astronaut mother blinked twice,
Then smiled and said, "That's so nice."
But they all got along,
Happily singing sweet songs,
And everything was sugar and spice.

Riding out to the launchpad in the Astrovan in the dark, early hours of morning induced no great philosophical thoughts. The crew participated in a little subdued, light banter, and Kevin commented on how excited he was to be launching to a specific destination. Mostly, we were all concentrating on being comfortable in our orange launch-and-entry suits. I was riding with my fingers stuck into the neck dam of the suit, pulling it out slightly from my neck so that I would not feel like I was choking. At the last stop before the launchpad, Dave Leestma, my boss, the head of the Flight Crew Operations Directorate, stood up to get out and say goodbye. Laughingly, I asked him if he thought he would still be my boss when I returned after my anticipated four-month stay on Mir. Laughing back, he replied that he sure hoped so, and then, as our eyes locked for an instant, I allowed myself to think of the unknown black void into which I was preparing to step. Then Dave said, "God bless," wrapping me in the assurance that my God

Lucid

and his would be with me and that our God knew the way through all the voids in the universe. I would never be alone.

We launched on time. As was usual on a shuttle flight, the crew was busy every waking hour—Linda Godwin and Rich Clifford checking out their extravehicular activity (EVA) suits in anticipation of their spacewalk, the first American astronauts would do on a foreign spacecraft; Ron Sega activating SPACEHAB and all of the experiments that were there; and Kevin Chilton and Rick Searfoss getting ready for our rendezvous with Mir.

The sounds of a loud trumpet playing "When the Roll Is Called Up Yonder" reverberated throughout the shuttle early in the morning of the third day. Startled awake by the music from mission control, the heads of the STS-76 crew members popped out of their sleeping bags. "Hey, Shannon, wake up," they shouted. "This song is for you! It is rendezvous morning!" My husband and daughters had selected the wake-up music for that day. It is my favorite song and the official "Lucid family song."

Years ago, I had the idea that every family needed a family song. In rebellion because "no other families have family songs," my kids and husband refused to submit nominations. By default, my song won. The capcoms hesitated to play this selection, not wanting to tempt fate, but my daughter and husband insisted that the symbolism was appropriate and that it was in no way suggesting a less-than-favorable outcome.

Just after the shuttle completed docking with Mir, the entire STS-76 crew crowded into the tunnel adapter while Kevin worked the mechanism to open the hatch, the last barrier separating us from the Russian space outpost. With a small sigh, the hatch floated down on its hinges, and with that, Mir and the shuttle became the only human habitation in the entire universe not restricted to the confines of Earth. Yuri Onufriyenko and Kevin Chilton, the two commanders, grabbed each other in a huge bear hug, pounding each other on their backs as they gleefully tugged at each other to see who would be the first to pull the other into their vehicle. The STS-76 crew surged into the Mir docking adapter, where Yuri Usachov, the Mir board engineer, also became engulfed in the gigantic hug. I trailed in last. After all, Mir was now my home, not just a place to visit. Then it was my turn to be included in the Yuris' hugs and smiles.

One of the STS-76 crew grabbed the mike and cord that Yuri had prepositioned in the docking module. Then we were all off floating,

Tumbleweed

jostling, single file, playing follow the leader, with Yuri and Yuri doubling as leaders and proud hosts. We all gathered around the table in the Base Block, positioned the camera, and had an official welcome broadcast to both the Russian and American mission controls. Business finished, we settled in to have a grand time enjoying each other's company, and I started to become a Mir 21 crewmate and not an STS-76 crewmate. Mentally, I exchanged Kevin for Yuri as my commander. It was a clean psychological break, even though we would continue to work and play together for several more days. All too soon, the STS-76 crew floated back to the shuttle. (Houston mission control always got very nervous when the shuttle was vacated for longer than ten minutes.) As I drifted in the Base Block between Yuri and Yuri and watched my "former" crewmates float back single file to the shuttle, I felt a tinge of regret that my time as a member of their crew had been so short.

The first order of business was for me to become an official member of the Mir 21 crew. For this to happen, my Soyuz seat liner and Russian launch-and-entry suit had to be brought over to the Soyuz. I also had to put on my suit and do a pressure check. A gigantic parcel consisting of my Soyuz seat liner, launch-and-entry suit, and survival gear that had to be placed in the Soyuz was transferred from the shuttle to the Mir. Then Yuri Usachov and I set to work. Back in Star City, John and I had marveled at how this was all packed together. Together, we wondered how we would ever remember how to unpack it, get our personal weights for CG (center of gravity) installed in the Soyuz, get the seat liner in, and do a pressure check. I should not have given it a second thought! Yuri Usachov had the package unpacked, my seat liner in, and me donning my pressure suit before I had time to figure out how to undo the fasteners! Dressed in my pressure suit, I squeezed into my seat on the right side of the Soyuz. As I pulled the seat belt and the kneecap protectors tight, I kept thinking about how small the real Soyuz felt. The mock-up back in Star City had not seemed so cramped! Yuri and I quickly did the pressure check, and everything was great, so I got out of the suit and left everything in position in the Soyuz so it would be ready at a moment's notice if there was an emergency. At the next comm pass, we informed Moscow mission control that we had successfully completed everything. Now, in actuality, I was an official Mir crew member!

Lucid

After Yuri Usachov and I finished the suit check, Yuri Onufriyenko asked me to come with him. He took me into Kvant 2 and showed me the panel that controlled the toilet. He explained just what I needed to do in case I had forgotten my Star City lessons. Then we floated into Spectra, where he and Yuri had prepositioned my sleeping bag and other personal equipment that had arrived on the last Progress. Finally, he handed me an optical-quality mirror that at one time had been part of some scientific equipment, commenting, "We thought that maybe you would like a mirror since you are a woman." I was touched by their thoughtfulness and thanked him for it.

The STS-76 crew was in the shuttle, Yuri and Yuri were in the Base Block, and I was in Spectra, settling in. The one thought that kept running through my mind was, "Where are the books that STS-74 brought up for me?" Are they really here? How could I nicely ask the Yuris about them without appearing to be dissatisfied with the wonderful way they had welcomed me? But I really wanted those books. Books just make anyplace feel like home. And I wanted to make sure that they were actually on Mir!

Before I figured out how to ask, Yuri Onufriyenko floated into Spectra, steering a large white NASA transfer bag in front of him. He grinned and said, "I have been keeping this behind the panel for you. We found it on Mir when we arrived. It must have come up on STS-74." Barely able to restrain myself from ripping the bag open, I looked through my "surprise." Not only were there books, but there was a zippered bag of lemon drops from an STS-74 crew member, Bill McArthur. Bill and I had flown together on STS-58, his first flight, and as we were being strapped in for that flight, I had given him a lemon drop. (Ever since my first launch, I had always brought lemon drops to the pad with me to suck on while waiting for liftoff.) This comforting personal thought that extended across the flights filled me with a warm glow. Lemon drops and books! What more could a person want? With that, I unrolled my new sleeping bag, floated into it, turned off the lights, and immediately fell asleep. It had been one long day!

COMBINED OPS AND GOODBYE

The next day, our "blended" space family began living and working together. The Mir 21 crew was awake the same hours as mission control in Moscow, but the STS-76 crew was sleep-shifting in preparation for a daylight landing in a few days, so we were on slightly different day-night cycles. In other words, the shuttle crew and the Mir crew went to bed and woke up at different times. This did not pose a problem; it just meant that in the time we were awake together, we had to work fast and furiously because we had thousands of pounds of supplies and equipment to transfer.

Floating across the hatch connecting the shuttle and Mir was like crossing the border between two different worlds. In the shuttle, the crew was busily making every second count—after all, time was counting down to a landing on Earth in a few days. Inside Mir, though, we were living a relatively normal life. Unlike the world of the shuttle, there was no frenzy to accomplish everything immediately. If we didn't finish a particular task today, there was always tomorrow. After that? Well, there would still be another tomorrow, stretching out to some dim time months ahead when we would finally need to think of returning home. Despite the presence of the shuttle crew, the three of us on Mir had our meals together, and between transferring bags of equipment, we took leisurely tea breaks together. We lived our lives at a normal pace.

Lucid

As an American member of the Mir crew, I was occasionally caught between these two worlds. I was the unofficial go-between, the ambassador. For instance, Kevin wanted the fertilized quail eggs, a science experiment to determine the effect of radiation on embryo development, transferred over to Mir immediately. Mission control in Houston was fixated on these eggs, asking in just about every radio transmission if they had been transferred yet. They could not be transferred until the Mir incubator was up and running. Each time I was in the shuttle, Kevin would ask me if the incubator was ready. Wanting to be as helpful as possible, I would quickly fly over to the Base Block, where Yuri and Yuri were working on the incubator. When I asked about the status of the incubator, they would look up in surprise, as if they were confused about my reason for asking. Their consistent reply was that it would be ready in due time—and it was.

Finally, the day came when all the food, science equipment, and shuttle-produced water had been transferred over to Mir. Rich and Linda had successfully done their shuttle-based extravehicular activity (EVA), and it was time for goodbyes. Suddenly, everything was a mad rush. I quickly gave emotional hugs to each member of the STS-76 crew. Kevin squeezed in one last gigantic hug before I helped Yuri Onufriyenko close the hatch.

I left the tunnel adapter and went into the Base Block to get on the radio and act as the translator during the undocking checklist for the English-speaking shuttle crew and the Russian-speaking Mir crew. Ever so slowly, literally inching, the shuttle crept away from Mir. I looked out the window in Yuri Onufriyenko's sleeping compartment into the overhead windows of the shuttle and saw the tops of the heads of my former crewmates. I had not realized how bare the tops of their heads were getting! It is amazing what a top-down view does for your perspective of someone else!

As we drifted farther and farther apart, high above the Tibetan Plateau, I felt just a little sad for the STS-76 crew because their adventure would be finished in just a few short days, whereas mine had only begun. My moment of introspection was interrupted by Yuri Onufriyenko's announcement that it was time for a party to celebrate the official start of the complete Mir 21 crew working and living together. We unpacked the grapefruit that had been a gift from the STS-76 crew and peeled and ate it, along with some of Yuri Usachov's personal cheese. We finished our celebration with a couple of hours of

Tumbleweed

sleep because ground had decreed that we needed to rest to get back to our normal day-night cycle.

I spent the rest of that first day floating to the window in Kvant 2 during each night pass and looking out at the shuttle, first as a bright and shining star, then as a dimmer and dimmer star as the distance between us increased. I remembered the letters I had hastily scribbled to my family and flight surgeon on the back of some flight procedure pages. I wanted to assure them in a private communication that I was safely aboard Mir, that I was fine, and that I knew everything would be all right. Thinking about those letters headed home by "space post" gave me a great feeling of satisfaction.

Perhaps I would have been less optimistic about letters sent by "space mail" if I had known that they would take four months to arrive. I assumed my family had received them soon after the shuttle touched down, but during an evening comm pass four months into my stay on Mir, Gaylen, my flight surgeon, said, "Thanks for the letter you sent on STS-76. I got it today." So much for postal service by space express. Give me UPS any day!

"COSMIC" LANGUAGE

Just as language spread its long shadow over every aspect of training, the language shadow penetrated every aspect of my daily life in orbit. In any discussion about day-to-day living on Mir, the first question is always, "But what language did you speak?" It always comes as a shock when the answer is, "Well, Russian, of course. After all, Mir is a Russian space station!"

Yuri and Yuri did not speak English. Sometimes at breakfast, Yuri Usachov would sing the one English phrase he knew: "Are you sleeping, are you sleeping, Brother John, Brother John?" (This was as much as his young daughter, who was studying English in school, had been able to teach him before he launched.) The very first day I spent on Mir and every day thereafter, I thanked myself for taking the time in Star City to have the evening language sessions with Nina. Those extra hours were my life preserver that kept me afloat, linguistically speaking, those first few weeks on Mir.

A few weeks before launch, it seemed that the Russians suddenly woke up to the fact that NASA really was planning to fly me to their station. After that, it seemed that I could do nothing right. It was as though the Russians were just finding fault with me and picking at me for everything, especially the way I spoke Russian. And I had to agree with the critics.

Tumbleweed

To me, it seemed that when it came to anything to do with the Russian language, my brain and tongue had become totally disconnected, and at times, I even thought that my brain had gone into a suspended state. I felt that where language was concerned, I had suddenly been transported back to ground zero—not that I had very far to go! Anything spoken to me was just so much gibberish, and unfortunately, anything that I tried to utter was also just so much gibberish. I really wasn't worried about this. I figured it was just a little mental overload and that as soon as I got in orbit, everything I had learned, and more, would come flowing out of my mouth.

Although I was not worried, it was still a tad embarrassing. One afternoon, just before the launch of Yuri and Yuri, the joint schedule for John and me showed us having a meeting in Building One. Still naïve to the end, we thought, "Oh, how nice—Russian management wants us to debrief our training in Star City, similar to what a shuttle crew does with NASA management just prior to a launch." As usual, we were wrong. It was a ceremonial function in the big conference room—the same one in which we had our initial welcome. Gathered there were many of our instructors and other assorted people from Star City, all decked out in their military uniforms or Sunday going-to-a-meeting clothes! Once again, John and I, garbed in our practical hiking-through-snowdrifts clothes, were very underdressed for the occasion. Some of the NASA management people, who were stationed at the embassy in Moscow, were also there.

We did not have a clue as to what was going to happen. We rapidly figured out that it was not the intimate debriefing that we had thought it would be! It turned out to be a time to list all the shortcomings of the crew. I listened to Russian management expound on the reasons why I should not fly on Mir and then listened to a cosmonaut who stood up and said that it was not right that I should be flying with the title "board engineer two." At that point, Yuri Onufriyenko just leaned over to me and said, "Don't listen. Everything will work out great in orbit." Finally, the head of training, with great reluctance, pronounced us ready, trained, and certified to go to Baikonur.

Then each member of the crew stood up on cue. I just did what Yuri whispered to me to do and said something in Russian. I could not think of anything to say even if I had been speaking English, much less Russian, but my mouth obligingly fell open, and noise came out. From the looks on the faces in the audience, I assumed that I was not speaking comprehensible Russian. Quickly reaching the conclusion that I had made enough noise, I closed my mouth and sat down. I told myself not to worry; my brain would normalize as soon as it got to orbit. And it did.

I love to talk. To me, it is a much more pleasant recreational activity than playing games or watching movies. For me, the definition of a good conversation is one that is punctuated by a lot of joking laughter. Prior to arriving on Mir, my one big dread had been the thought of living there for a four-month planned stay without being able to laugh with my crewmates. In other words, I was concerned about not being able to joke around with words and situations like I normally do with friends and acquaintances.

The second day that I was part of the Mir crew, we were floating around the table, and I started to kid around with Yuri and Yuri. They both spent the entire meal laughing with me, not at me or my Russian. I breathed a huge sigh of relief. We were going to have a great time living on Mir together! I was home free!

There was never a problem with communication among the three of us in orbit. Obviously, I had not received the gift of tongues and was not speaking colloquial Russian like a native, but there simply was never a problem with communication despite my own deficiencies with the language. A large part of this was due to the consideration and effort that Yuri and Yuri put into making me feel a part of the crew. Again, I must admit, I love to talk, and from the minute STS-76 undocked and until STS-79 docked, I rattled on in words that I considered Russian.

When the three of us were together, Yuri and Yuri never conversed in Russian at breakneck speed, leaving me unable to follow the conversation. They were always very conscientious about making sure that I was included. Yuri Usachov, especially, had a real gift for what a linguist would call restorative abilities. What that means in laypeople's terms is that no matter how I butchered the grammar or the pronunciation, Yuri could generally figure out what I was saying. He was also good about first introducing a new conversation topic with a simple sentence and then making sure that I was following along. We all quickly learned to compensate for the many language shortcomings I had and did not let them stand in the way of having a good time together.

Tumbleweed

We actually had fun with the language. Russian is dependent on verbs. Unfortunately, my Russian vocabulary was limited in this part of speech, so I experimented a little. I quickly discovered that if I just used an English verb, changed the pronunciation so that it had a Russian sound, and tacked on the appropriate Russian ending, I sometimes hit a home run; it was a verb that Yuri and Yuri understood. Later, I became a little bolder, and if my newly invented word elicited only polite blank stares, I would say, "Come on, you guys. Surely you know that word! It is really pretty bad when I have to start teaching you how to speak Russian!" They would laugh and add my newly formed word to their vocabularies. Soon there was an entire list of words that only three people in the universe spoke and understood.

We constantly joked about our "cosmic language." I did not realize how entrenched these words were in our daily life until the exchange crew arrived. I overheard Yuri and Yuri laughing with the new team members, Valeri Korzun and Sasha Kaleri, saying, "These are the words that you need to know to live and work with Shannon." Then they recited the entire litany. Valeri and Sasha immediately started to use the "Shannon list." The first few times they used words from the list, it was as a joke, but the terms soon became second nature to them also. After the flight, we still sometimes laughingly used them when we were together. It was the "insider" thing to do!

Another reason language did not become a barrier was because neither of the Yuris ever acted like it was a burden to listen to me making mincemeat of their language, and they never bothered to correct me. I would have thought that my constant jabbering would have gotten on their nerves at times, and I was certainly prepared to try to be quiet if asked (although it would have been a great burden because I really do like to talk). One morning, I had been silently wondering about where on earth—or I guess I should say, where on Mir—I could possibly look to find the missing piece of equipment I needed for the day's work. As I floated by deep in thought, Yuri asked me if something was the matter.

I replied that I was fine. Yuri explained his question by saying that because I hadn't laughed all morning, he thought something must be wrong. That made me feel good. Not only did I realize that he cared about me enjoying the day but also that he was missing my laughing banter!

Lucid

Before the launch, I made sure that I had an English-Russian dictionary to bring over with me to Mir. One of the first things I did on Mir was to Velcro this dictionary above the table in the Base Block, guessing that might be the place I would use it the most. Surprisingly, we only used it once during the entire six months I was on Mir. One lazy afternoon, we got into a religious discussion. Yuri was trying to explain a religious holiday celebrated in the Russian Orthodox Church. We got stuck trying to understand each other and pulled out the dictionary. It turned out to be of no help at all. Every other time, if I was not getting my point across due to my limited vocabulary, I just sort of backed up and said the same thing a different way. The Yuris did the same thing when I did not understand them. It was much faster than trying to look up words!

One night as we were floating around the table in the Base Block, drinking tea and waiting for the last comm pass so we could call it a day and go to bed, Yuri Onufriyenko just out of the blue asked, "Shannon, you don't understand absolutely everything that we say, do you?"

I laughed to myself, sure that I didn't even understand half of what they said to me, much less "absolutely everything."

Yuri continued with his linguistic observations, commenting, "You speak much better Russian with your hands than you do with your mouth!"

Then we laughingly talked about the dominant role that body language played in our "cosmic language."

Toward the end of the flight, after we had been together some five months, this topic came up again. Yuri Usachov commented that we did not need words to understand each other, and we all laughed because this was so true. After spending so long together, just the three of us, we instinctively knew what each other wanted and was going to say before our lips and tongue could form the words. Just how much this had become a part of our lives was brought home to me a few days after Yuri and Yuri had departed and I was working and living with my new crew, Valeri and Sasha. One day, Valeri was positioned at the command post in the Base Block, and I was at the comm spot where I had listened to the ground on every comm pass for the last six months. I saw Valeri flick his eyebrow, so I flipped on all the TV downlink switches.

Valeri yelled, "Shannon, what are you doing?"

Tumbleweed

"I just turned on the TV—"

"No, no, we don't want the TV on; turn it off."

Feeling hurt, I quickly flipped all the switches off, and then I really laughed because I realized that I had just responded to the "Skiffer cosmic body language." The only problem was that no message to turn on the TVs had been sent. Valeri was not part of the Skiffer crew and did not know the intuitive body language that had been developed over the last five months. I laughingly explained to Valeri and Sasha that because we were now the "Frigate" crew—the call sign chosen by Valeri—it was time to start working on developing the Frigate cosmic language!

COMM AND ALL THAT

Our day-to-day living on Mir was bounded and structured by comm. The first and the last scheduled activity of each day was a comm pass. Unlike the space shuttle, which almost always had the capability of talking with the ground via several Tracking and Data Relay Satellite System (TDRSS) communication satellites, Mir, in general, only used Russian ground sites for daily comm. This meant that we could talk with the ground only when it passed over a Russian ground station. On average, this happened every orbit, or about every ninety minutes. Each pass generally lasted about ten minutes. The Russians did have communication satellites similar to the US TDRSS satellites, called Altair, but it was expensive to use them. As a result, they were only rarely used to communicate with Mir. On the occasions when Altair was used, we were able to talk to the ground for an extended period—sometimes up to forty-five minutes. Because of the extended communication time and the reliability of the contact, the family videoconferences, psychiatric support conferences, and press conferences were generally scheduled using Altair.

Work Talk
Our commander wanted each of us to be on comm during every ninety-minute pass. As he explained, our capability for talking to the ground was limited, unlike that on the shuttle; therefore, we needed to be ready if and when the ground needed to talk to each of us. Not

Tumbleweed

being prepared for each comm pass would result in wasting valuable comm time or, perhaps, missing an entire opportunity. Every evening, the next day's scheduled comm passes were uplinked to us. Before going to bed for the night, I would always copy out the comm pass times in my notebook so I would have the anticipated times ready for the next day.

I looked forward to these comm passes because they provided small social breaks throughout the day. No matter how occupied we were with our various individual timeline tasks, we would have this built-in break when we would all gather in the Base Block. Prior to and just after the comm pass, we would catch up on how each of our various tasks was progressing. Waiting for the comm pass to occur also provided time for the three of us to crack a few jokes—usually at the expense of the ground.

For the Mir missions, the US provided two ground stations located in the United States: Wallops and Dryden. In theory, these two sites were to provide increased options for comm when an American was on board. In actuality, there was no point in wasting time when those two stations were scheduled because we never heard anything except static. It became a standing joke between us that because it was "American comm," I would be the one who had to don the headset and spend ten futile minutes waiting for ground contact. Only rarely did we hear the Russian control center through these stations, but many times, we picked up various emergency bands through the Wallops station. We listened with real interest to such Earth events as an ambulance driver ordering pizza to feed the two lost children he had just picked up.

The NASA contract with the Russians stipulated that there would be two comm passes every day so that a NASA person stationed in the Russian control center could chat with the American aboard Mir. Once in the morning and once in the late afternoon, I talked with either Bill Gerstenmaier, the lead NASA person assigned to the Russian control center for the NASA-2/Mir-21 flight to oversee the planning and executing of all the NASA experiments aboard Mir, or Gaylen Johnson, my NASA flight surgeon for this flight. Bill and Gaylen worked together to coordinate with the Russian flight planners to timeline the various NASA experiments. Bill also worked with the principal investigators for the science experiments. These investigators were located primarily in the United States, although some actually

Lucid

came to the Russian control center when they thought their experiments would be carried out. They did this to ensure that the procedures were correct, as well as to be available to answer any questions I might have or figure out ways to work around any anomalies.

Gaylen's primary tasks were to make sure that I stayed in good health and to work with the Russian medical team. He also had to keep track of the environmental parameters to ensure that I was living in a healthy place. With all this help, I did not have to do any worrying. It was all being done for me!

I looked forward to my NASA comm passes. Frankly, it was good to speak a little English, and it was always great to talk with Gaylen and Bill. Without exception, every comm pass with them was positive and upbeat. On Mir, my world became very narrowly focused. From my perspective, the three of us—Gaylen, Bill, and I—were the only NASA personnel who knew what was going on and the only ones working this mission. I got into the habit of thinking about us as the Three Musketeers—one for all and all for one. We were a team. But I was wrong in thinking it was just the three of us working the mission, and that fact became very apparent to me when, after my flight, my job was to be the astronaut office representative to the Phase 1 office. Only then did I realize how many dedicated people were working very long hours, many times at great personal sacrifice, to make each and every one of the Phase 1 missions a success.

At the conclusion of a comm pass one day, Yuri Onufriyenko removed his headset and asked me, "Shannon, how many people does NASA have working down in the control center?"

I shrugged my shoulders and said, "I don't know, maybe five or six. Why?"

"Because every time you ask Bill a question, he always gets back to you with the answer by the next comm pass!"

I knew that Bill and Gaylen were doing a great job, but it still sort of caught me by surprise that Yuri, who spoke and understood no English, had been able to pick up on how great the NASA support team in the control center really was. At that moment, I could not have been prouder to be part of the NASA team!

Tumbleweed

Calls from Home

The Russian plan for Mir crews maintaining contact with their families while they were in orbit was that every week, usually on Saturday or Sunday, all crew members had a dedicated comm pass to talk with their families. I always marveled that my husband and children, sitting in our family room in Houston and using our home telephone, could be patched into the Russian control center and then from there be patched up to Mir. They did not even have to be all together in one place. Several times, my son, who at that time was trapping wolves in Minnesota on his summer break from Texas Tech, was patched into the conversation. Another time, I was talking to my son and I heard the distinct roar of interstate traffic. When questioned about it, he said, "Oh yes, I'm headed back to school, and I noticed it was time for our family call. I just pulled into this Shell station and dialed home. So here I am, in a phone booth, in the blowing sand in the middle of the Texas Panhandle, talking to my mama who is two hundred miles above Earth, orbiting over Russia."

On alternate weekends, we had family videoconferences instead of a phone call. Generally, Yuri and Yuri had theirs together. Both of their families traveled to the Russian control center, and then, via the Russian comm satellite Altair, their images were linked to our small black-and-white onboard Sony monitor, and simultaneously, the image of Yuri and Yuri would be sent down to the control center to their families. Many times, Yuri and Yuri asked me to be part of their conference. I found it very poignant to see how their children changed over the course of our flight. Yuri Onufriyenko's youngest son went from being a bouncing preschooler to a young man ready for kindergarten.

For my videoconference, my family would go into the videoconferencing room at the Johnson Space Center. Because these events always happened on the weekend, this meant that JSC employees had to work overtime just so we would be able to talk and see each other. The dedicated comm personnel always did a wonderful job making sure that the JSC end of the comm was up and working great.

My family and friends sat in a videoconference room equipped with a huge screen on which they could see me and another screen on which they could see themselves. After the first family conference, I was really wishing that someone would give them a quick course in videoconferencing! One of my daughters, who will remain nameless, spent most of the time looking at her image on the screen and then adjusting her hair more to her liking. My husband always sat with his arms folded across his chest and a frown on his face, as though he was daring someone to interrupt the session. My son-in-law was always there along with my other daughter, but more into gadgets than words, he managed the controls. He spent the whole time zooming the image in or out, panning the entire room from the ceiling to the floor. He would zoom in on someone's fingernail or the tip of a nose. Sometimes the camera would catch him flitting from control to control as he raced around the room, making sure that all the technology worked and that he used every bit of it!

As I mentioned, up on Mir we received the images on a tiny, off-the-shelf black-and-white Sony monitor. The picture quality was marginal. I would squint and sometimes put my face almost on the screen, trying to figure out who was pictured there. Another huge drawback was that the audio was often atrocious! The echoing was just horrendous. That is, when there was audio. Many times, it seemed that one or the other end would not be able to receive the audio. And then the event became a game of twenty questions: "OK, now shake your head yes if you hear me. Shake your head harder; I can hardly see any movement. Try holding up both hands if you hear me. Great! You can hear me. I will just talk. Put your hands down if you lose my voice. This week . . ." And then I would drone on in a monologue of the week's events.

The sessions where video was received and transmitted but there was no audio at either end were the most challenging. The first time this happened, my daughters instantly rose to the occasion. They grabbed scrap paper and a felt-tip pen and started writing questions, which they held up to the monitor. I replied back in the same manner. Yuri and Yuri, who many times slid down under the table in the Base Block so that they could unobtrusively watch my family trying to communicate with me (they loved Jeff's antics and named him the family clown), thought this showed great ingenuity on the part of my family. From that time on, whenever I was scheduled for a family videoconference, they would always scrounge up some paper and a felt-tip pen for me, "just in case."

My favorite family conference was on my son's birthday. Before the flight, Bill Gerstenmaier had asked me which dates I considered

Tumbleweed

important family days that would occur while I was in orbit. He said he would try to arrange a family conference for me on each of those days. Originally, my list included only Mother's Day, my mother's birthday, and Father's Day. Due to Bill's negotiations, I had great phone calls with my mother and father on all these days. I think it slightly overwhelmed my mother to be receiving a space phone call on her eighty-second birthday. She was truly at a loss for words. Knowing her, I could just hear her mind clicking away and thinking, "This is not what I bargained for the day I gave birth to you!"

The unanticipated delay of my return flight meant that I would miss both my son's and daughter's birthdays. When informed that I would be in orbit longer than planned, I gave Bill the dates of their birthdays so that he could start working on trying to set up videoconferences. As soon as I mentioned that to him, he replied in his typical can-do Bill fashion that he was sure he could arrange videoconferences for those dates.

Now it is a mother's true confession time. I knew that Michael was going to be twenty-one on his birthday. After all, I was there twenty-one years ago! I got it in my head that this was a really special birthday because the Mir 21 crew could wish him a happy birthday on his twenty-first birthday on August 21. I loved this triangle of twenty-ones. It just made the day feel so special. I had it all planned out. I took three pieces of paper and wrote:

Happy Birthday, Michael,

On your 21st birthday on August 21st,

Love, the MIR 21 crew.

I decorated the sheets using my colored markers and then drew borders of flowers. (I know, flowers are not really manly, but that is all I can draw.) Yuri and Yuri laughingly agreed to participate in my "space" birthday card. Everything was ready. I anxiously scanned the next day's timeline on the evening of August 20. Nothing about a family conference. No note for such an event was mentioned for the next day. I felt devastated. Bill had never failed me before! I couldn't ask him about why there was no time to talk to my son on his birthday because I knew that he had tried his best, and I didn't want him to feel bad. It was bad enough that I felt so devastated. Well, somehow, I got through the next day, August 21, feeling sorry for myself, thinking about the family all together without me. That evening, I saw the timeline for the next day, August 22. Printed out on the form was a

Lucid

chunk of time for a "family conference for Shannon." It wasn't just a phone conference but a videoconference! Thirty minutes' worth! Life was once again golden. So what if they were a day off? Bill had done his best and come through like I knew he would.

The next morning, the video from the conference room at JSC came in almost crystal clear. It was the best reception I had ever seen on Mir. My three children and my one son-in-law were all sitting in a row, grinning from ear to ear. My first words—after the "happy birthday," of course—were, "Well, did you have a great birthday yesterday, Michael?"

Looking a little puzzled, he replied, "Yesterday? No. But I am today because today is my birthday."

"Today is your birthday? But isn't today August 22?"

"Yes, it is, and that is my birthday. You know, the day I was born."

"Are you sure you were born on the twenty-second? I thought you were born on the twenty-first?"

"You are really out of it! You can't even remember your favorite baby boy's birthday! Are you sure you're my mother?"

All that time yesterday feeling sorry for myself was wasted! I had been so sure!

Michael then said, "Well, since you can't be here with us, I brought the birthday celebration to you," and he put a cooler up on the table and proceeded to open it, taking out an ice-cold Coke. After a few dramatic sips, he said, "Oh, that tastes so good. There is nothing like a Coke to satisfy a thirst! Here, let me offer you some." He held the can up to the camera. Then he put a white sack on the table and proceeded to lay out a few gooey doughnuts. "Let me see . . . which one will I choose this morning? It is always so hard to decide! Should it be the cream-filled chocolate one or the apple fritter? Oh, I forgot about you, poor, poor Mother. You can't share one with me! You have been without doughnuts and Cokes for six months! Well, let me tell you how good they taste!" After finishing a doughnut, he proceeded to lick his fingers. "Well, you know, Mom, if you would ever decide to come home, maybe we would get you some doughnuts also!"

After that, the "sight and sound" birthday card that Yuri and Yuri and I sent down seemed a little anticlimactic. Then the Yuris wanted to see more of Michael's wolf pictures. All too soon, the thirty minutes were gone. And yes, it had been a great birthday celebration. Later, when I was home, the girls said how they also really enjoyed that conference and that it was their favorite of all our Mir family conferences. As they left the room that morning, there was a crowd of high-level NASA managers waiting to have access to the room. Kawai said, "You know, that just made me feel really special that they all had to wait outside until our Earth-to-space birthday party was finished. The NASA folks had their priorities right that day!"

Press Conferences

We were always elated to see a family videoconference scheduled, but the same could not be said for the times when we saw a press conference scheduled. It was hard speaking to the media when we had not seen any news for months, when we had been free-floating in our own little world and had no connections with the humanity outside our tin can orbiting above Earth. We were living without feedback. It was like sending signals into a vacuum and never getting any return. We had no idea how what we were saying was being perceived by people on Earth below. And of course, there were the technical problems. Many times, there was such a loud echo that it was almost impossible to make out the words, and to make matters worse for me, sometimes the reporters were speaking Russian and expected Russian answers! It is hard enough answering questions when you can hear every word clearly, but when the words are garbled and echoed back at you, it is truly an impossible task.

We generally managed to have fun with them. Invariably, one of us would start to joke around before comm was established, and then we had a hard time not letting the joking atmosphere leak into the interview. Other times, video could not be established, and the session turned into just voice. Once, there was supposed to be video, and then I was told that due to technical difficulties, it was voice only. Therefore, while I was answering the questions, I had my toes wrapped around a large bungee cord that was attached from the floor to the ceiling and was "walking" up and down it—yes, like a tightrope walker in the circus. (I loved doing this when we were on comm. I loved the image of me being a tightrope walker, and also, I just liked to be moving around. I never realized just how much time I spent moving until one day Yuri Onufriyenko said, "Shannon, why are you always moving? Can't you ever just be still?" I did try to perfect a stationary float after that, but many times I backslid because I really do like to move.) After the event, Gaylen mentioned something about the fact that, in his

Lucid

professional opinion, I was still looking good, and I, somewhat shocked, said, "That news conference had a video downlink?"

He answered, "Oh yes, they got the video link established a short while into the interview."

I have always wondered just what was downlinked that day!

Once during a planned downlink, I decided to play a practical joke on Gaylen. The entire time we were training in Star City, the flight surgeon was required to be present as an observer for each simulation in the Soyuz. So every time I was in the suit-changing room donning the launch-and-entry training suit, Gaylen was there, watching this exercise. During the winter, it was very cold and very dry in Star City. This dryness meant that I used lots of lotion on my dry, flaking skin. It also meant that my fingers were cracked around my fingernails. One day, because I had an uncomfortable crack in one of my fingers and because I was afraid that the sharp piece of skin that resulted from this crack might poke a hole in the suit, I put a Band-Aid around the tip of my finger as a precautionary measure. At the scheduled time, I donned my suit under the watchful eye of Gaylen. The Russian flight doc called him out of the room, and soon after that, Gaylen came back in, motioned me aside, and asked in a low voice, "Shannon, what is wrong with your finger?"

"My finger, what do you mean?"

"The Band-Aid—why do you have a Band-Aid on?"

"Oh, that? That's just so that the sharp piece of dry skin on my fingertip won't poke a hole in the suit."

"OK," he replied.

Later, as we left the sim together under the cold, black winter sky and approached the point where our paths diverged, I asked Gaylen why he wanted to know about the Band-Aid. He replied that as soon as I showed up, he had been asked what was wrong with my finger. He was yelled at and told that he was a very poor flight surgeon because he didn't care for the health of the crew member he was in charge of and, in general, made to feel bad because he had not asked about the Band-Aid's purpose or, maybe even worse, hadn't noticed it.

It was time to see if my flight surgeon had learned his lesson. On the last Progress, one of the many cards from my daughters had a few fluorescent Band-Aids tucked in the envelope. (Don't ask me why. My guess is that they just happened to see them when they were rushing around trying to meet the deadline for turning in some mail to get over to Russia to get on the Progress. And admittedly, I do have an addiction to all things fluorescent.)

Just before the time for the press conference, I quickly ran a brush through my hair, although it would help me look presentable for only about fifteen minutes, and wrapped a bright fluorescent-orange Band-Aid around the middle finger of my left hand. All during the press conference, I made sure that finger was visible without trying to bring obvious attention to it. The press conference ended, and the waiting began. Finally, several orbits later, it was time for a NASA comm pass. Gaylen came on the comm and, in a sort of tentative voice, said, "Shannon, what is wrong with your finger?"

I couldn't help it. I literally doubled up (very easy to do in space) and started to laugh and laugh. I was laughing so hard I could hardly choke out the reply, "Well, I guess you have passed with flying colors, so to speak, your remedial flight surgeon test." There were a few seconds of dead time before understanding dawned and Gaylen also started to laugh. By the time we got ourselves under control, there was not much time left for an update on the day's work.

Ham Radio

Before launch, wanting to make sure that I would have someone to talk to on the ham radio that was located on Mir, I bribed my family by offering to buy a small ham radio for any one of them who would study and take the test to get their ham radio license. Shani and Jeff took me up on the offer. I was so glad that they did. Being able to talk to them this way was one of the best means of communication. The voice quality over the ham radio was excellent. It was always clear and understandable. Best of all, the ham radio could be used without any official coordination. All Shani, Jeff, and I needed to know was when Mir was passing over Houston. After the first few contact attempts using their small mobile set, where we were only able to get in a minute or two of talking, they joined the JSC ham radio club and were able to use the club's ham radio at the center. It was a great setup. It had a computer-controlled antenna that was able to track the signal from Mir. This allowed us to remain in contact for sometimes up to ten minutes. Every time Mir passed over Houston during daylight hours, Shani and Jeff were there, Shani doing the talking and Jeff manipulating the controls. Jeff worked on-site at JSC, and Shani was

Lucid

working just off-site, so it was very convenient for them to meet there and eat lunch together while waiting for signal acquisition from Mir.

Aboard Mir, I would watch the computer display of our orbital location. When we were approaching Houston, I would key the mike and say, "This is RO Mir. How do you copy?" Shani's voice would come in, and we would start talking about daily events, such as whether Mike was keeping the grass mowed, what to get Kawai for her birthday, and what news events were taking place down on Earth. The only constraint to our conversations, and what I constantly kept in mind, was that they were in no way private and that every word we were saying was being listened to by many other ham operators.

Sometimes Shani and Jeff were joined by some of my friends and coworkers from JSC. Ellen Baker was one of the few astronauts who understood the psychological importance of maintaining contact with the Mir crew, and she was almost always there. We had many pleasant conversations discussing what her daughters were doing in school and what books she and I had been reading.

I thought it was ironic that it was my "noncommunicative" child, Shani, upon whom I was dependent for news from planet Earth. When she and Kawai were in school together, I always knew what was happening because Kawai furnished the entire family with nonstop reporting of the day's events. When Kawai left for college, I told Shani that she had to really try to let us know what was happening in her daily life. One day, the mother of one of her friends told me how Shani and her friend had seen a girl passed out on the hood of a car in the school parking lot. Of course, Shani had never mentioned this to me. When I asked her about it, she said, "Oh yes, that happened," and then, when I asked why she had not mentioned it, she said, "Oh, you thought that was interesting?" This is just one small example of what I mean by not being a spontaneous communicator! So our first few sessions were filled with me prompting her to try to remember something that she had heard on the news or seen in the paper. But to give her credit, she did a great job. After the flight, when I sat down in my La-Z-Boy recliner and started going through my six-month stack of accumulated news magazines, I found only one major story that I had not been aware of. When the pressure was on, Shani came through like a real trooper!

At the end of every Houston ham radio pass, Yuri and Yuri would quiz me on what was happening down in Houston, down on planet

Earth. If Shani had told me some interesting news story, then we would discuss it all through the next meal. Ellen kept us abreast of astronaut office gossip as to what was going on with space station plans, and Shani told us about such things as the Olympic bombing and the crash of a TWA flight. All too quickly, the comm pass would be over, and we would lose the signal from Houston. Then we would return to our daily routine, refreshed from our casual brush with earthly matters, much like the refreshment given by the first gulp of ice water on a hot summer day.

I've Got Mail!

Put as bluntly as possible, email was what made my stay in Star City a pleasure and not a chore. Because of CompuServe, I had daily contact with my family. Each one sent me a note every single day, and in this way, I did not lose continuity with their lives. I knew when my daughter's car engine caught fire on the freeway, when my husband poured Windex into the washing machine instead of Clorox (each child took the time to describe this incident in detail—I did not hear of it from my husband!), and what the weather was like, so I was living their lives with them almost in real time. I knew that no matter what would happen on Mir, the intimacy that email could provide would keep me going.

Before my flight, we knew about the ham radio aboard Mir, and we also knew there was the possibility it could be used to downlink and uplink electronic messages. A smart person told me that, theoretically, messages that were in text files could be transmitted by the packet system. Prelaunch, I had someone show me how I could use the onboard computers to type messages in a text file. I had the instructions typed out and flew them in the medical support kit that I put in my flight suit pocket just prior to launch. Yes, it was medical— it was the key to my primary psychological support!

One of the last things I did in Star City before returning to the States for launch was to give Gaylen my laptop computer loaded with CompuServe. I explained to him how he could hook it into any Russian phone line—in the control center, in his hotel room, anywhere—and connect and pick up my emails from my family. I told him how his most important job as a flight surgeon would be to figure out how to uplink all of them to me from the Russian control center. I told my family to continue to send me emails like always, and Gaylen would

Lucid

get them up to me. To motivate Gaylen, I told him that until he figured it out, I would just dictate messages when we were on comm together, and he could type them up and send them to my family. For a couple of weeks, this is what he did. (After the flight, when I asked how they liked the messages from Gaylen, my daughters told me that they knew immediately when they started getting the messages I typed up because they were full of typos and misspelled words. None of Gaylen's messages ever had any misspellings!)

Gaylen soon figured out whom to talk to in the control center, and then he figured out how to get the CompuServe emails in the right format to send them up. My preflight intuition was right. These email messages kept me going day after day. They were what gave me immediacy with my family. Through the messages, I learned about new boyfriends, broken-down cars, changing college majors, changing philosophies of life, and more mundane things like what the neighbors were saying about the uncut lawn. Desperate to think of something to say every single night, Mike took to typing the printed messages from the cards that we received and describing the picture on the front of the card. And I loved it. Every single word was worth its weight in gold.

The highlight of any day was when the Russian control center would try to packet up messages. Of course, the work messages were first in the queue. And then, if there was time, the personal messages would come up. The best words that I could hear would be Yuri Usachov saying, "Shannon, you've got some mail." I would take a floppy disk, transfer my uplinked messages to it, and stick the floppy in one of my zippered pockets. At the end of the day, after saying good night, I would transfer the messages onto my laptop. I slowly read and savored each and every word.

When the system was working really well, I would receive messages at least twice a week. But then there were the long periods when the onboard packet system would not connect with the ground. I would try to be very patient and understanding, but then I would just have to finally ask either Bill or Gaylen if they understood what the problem was with the packet system. They never did. It seemed that there was only one person in the control center who could work the system, and sometimes he was on vacation. Other times, only one ground station would connect, and our orbit never seemed to go over that station. Sometimes there was not even a pretense of a reason.

Tumbleweed

When I began to think that I could stand waiting no longer, the chalk-grating crackle of the packet system would emit the sound of being connected. This was the modern technological equivalent of the far-off drumbeat of horse hooves on the prairie as the Pony Express was inbound to the way station. The crackle of the packet, the drumming of the hooves—sweeter music I had never heard! Mail was inbound!

The thing that surprised me the most about communications was that we learned about important events through the nonofficial channels first. Yuri and Yuri had their mission length increased from five months to six. We learned about this when I received an email note through CompuServe via the ham radio packet system from Tom Akers, the STS-79 crew person in charge of transfer. He wanted me to ask Yuri and Yuri some transfer questions because they would now be the crew present when STS-79 arrived. Only after I mentioned this note to Yuri and Yuri, and only after they queried the ground, did the ground acknowledge that, yes, their mission had indeed been extended.

I heard about the extension of my mission first from my daughter, who kept me informed of the latest NASA rumors via email. The final decision was also revealed to me by my daughter in this way. After NASA had held their official meeting and the decision was made to delay the launch of STS-79—my ride home—by six weeks, Shani and Jeff were out at the ham radio shack, along with some of my friends from the astronaut office. Before radio contact with Mir was established that day, there was a spirited discussion about whether they should tell me that my mission had been extended. They were discussing among themselves whether they needed approval from management to tell me. Shani listened for a few minutes and then said, "Oh, don't worry about it; Mom already knows. I already told her in an email message yesterday."

Now, I must give NASA credit. Frank Culbertson, the head of Phase 1, was trying very hard, through official channels, to inform me of the slip. But it was difficult to schedule time when he could talk with me over air to ground. He finally did get scheduled and told me officially that STS-79 and my return to Earth had been delayed. After this, every time Yuri and Yuri saw on the daily schedule that I had a comm session scheduled with Frank, they would say, "Shannon, bad news from NASA today—Frank wants to talk to you." And so it went.

Lucid

The biggest "noncommunication event" was the last-minute replacement of the exchange crew. Yuri and Yuri had been on Mir for six months, and it was time for them to head home for a very specific reason—the shelf life or space life of the Soyuz was about to be exceeded. The Soyuz, Yuri and Yuri's ride home, had a six-month guaranteed lifetime in space. A new crew was to arrive to replace them and their Soyuz. Along with the new Russian crew members would be Claudie Haigneré, a French cosmonaut researcher, who would spend two weeks on Mir and then go home with Yuri and Yuri, leaving me aboard with two new Russian crewmates.

It was Sunday, four days before the new crew was to join us. Their arrival was a big event in our lives, so we had been discussing the impending day for some time. Yuri Onufriyenko had a family telephone conference. As soon as he finished talking to his wife, he called to us and said, "Guess what? My wife told me that the gossip around Star City is that Gennady Manakov and Pavel Vinogradov are not coming. They were replaced yesterday by Valeri Korzun and Alexander Kaleri."

"What?" Yuri and I replied in unison. We had so many questions: The crew has been switched? But why? Why did no one tell us? What is the ground thinking? That we are going to open the hatch and see a new crew, and that is how we are supposed to find out? That we'd open the hatch, expecting to see Gennady and Pavel, but surprise— it's Valeri and Sasha? Was this any way to run a space program?

Our speculations were endless. I was glad I had a family telephone call scheduled. I could ask my husband what was going on. As soon as I heard Mike on the radio, I asked, without even asking how everyone in the family was getting along, "Have you heard about a crew swap for us?"

And he answered, "Oh yes, I read about it in the paper a couple of weeks ago. It was found out that Gennady had a heart problem, and so the backup crew will be launching in a few days!"

After my teleconference, I relayed to Yuri and Yuri what Mike had told me. Immediately, they got into a very spirited and dynamic Russian conversation that I found impossible to follow. This was the first and only time that they had a conversation in which I was obviously not included. After an extended period, their conversation slowed down to the point where I could join in. Of course, I had endless questions, for which they had no answers. They were in the

Tumbleweed

dark as to what had happened as much as I was. But they did have one advantage over me: they at least knew who the new crew was. True, I knew Valeri and had talked to him many times in Star City, but I did not have a clue as to whom Sasha was. Yuri and Yuri tried to describe him to me. They were sure that I must have seen him at least once, but if I had, I certainly could not place him. So, I met Sasha, a crewmate whom I ended up living in space with for a month, for the first time when the hatch opened. Oh well.

Later that evening, Yuri told the ground what we had heard and asked them just who we should expect when the hatch was opened. I thought to myself, "Thank goodness for our 'outside' methods of gleaning information, or we would never know anything important!"

FOOD AND MEALTIMES

Per the agreement between the Russian space agency and NASA, we had food from both countries: two-thirds provided by Russia and one-third by NASA. The food NASA provided was shuttle food, not anything specially prepared for a long-duration flight. It consisted of dehydrated food, which was ready to eat after adding hot water, and MREs, food in irradiated packets that would not spoil at room temperature. These are the same packets the armed forces use in the field.

The Russian food was similar. There were dehydrated packets of food, but they also had canned fish, the "little cans," and canned casseroles, the "big cans." The Russian food canisters also contained tubes of fruit juice. I did not like these because when we tried to use them, the paint on the tubes came off on our hands. Also, there was not enough fluid in them to be worth the effort. I had to drink five or six to quench a normal thirst.

Overall, the food was good. Once during the flight, the flight surgeon asked me which I preferred, the Russian or the American food. The Russian medical support people had prompted him to ask me. I answered truthfully that I liked both of them and that having both types gave us more variety. However, if I had to choose one thing that I liked the most, it would be the canned casseroles from Russia. These casseroles were generally different types of meat, such as beef or pork, and potatoes. I liked them because they were a little greasy

Tumbleweed

and hence the most like "normal food." The one thing that we had a real shortage of was a selection of vegetables. Invariably, the vegetables were the first things to go from a newly opened container. We always had plenty of meat cans left over.

My one criticism of the Russian food was that the portions were too large. Portion size was an important consideration because we had a sort of informal rule that what you opened, you ate. This was to cut down on any messy trash problems.

Russia and NASA both prepared menus for a seven-day repeated cycle. These seven-day increments were put in metal containers and sealed with duct tape. Taped to the inside lid of each canister was the seven-day menu. None of us followed it. We ate our favorite foods first each time we opened a new container. Sometimes we opened one container to share, and other times we each ate from our own containers for a week. Theoretically, they were packed with food that we had selected. In reality, we spent the majority of the flight using the unopened containers left on Mir from other flights. Because we used the oldest ones first, it wasn't until late in the trip that we started using the ones that had the food items we had selected.

I gradually took over the job of arranging the food containers. For instance, I would take all the small cans from the open containers and consolidate them into one. Likewise, I did the same thing with all the drinks—the tea and the coffee—and the desserts. This saved us a lot of time because when we had several containers open, we would have to rummage through them all just to find a bag of tea. I asked Yuri and Yuri why the ground did not pack them in this way to begin with. He answered that originally the food had been packaged like that, but crew members ate the good food and left what they did not like. Still, it was much more convenient to go to the container that I had labeled "drinks" when we needed a drink rather than having to open lots of containers and have packages float out as we searched for a drink.

At first, breakfast was the hardest meal for me to find something to eat. I did not like either the Russian or American breakfast foods. Instead, I decided to have a bag of Russian soup for breakfast. It was nice and hot and tasted great. So each morning, I had a bag of Russian soup and a bag of Russian juice. The Russian juice was more like dessert than juice. It was pulpy, thick, and sweet, which is great for a dessert but not wonderful as a thirst quencher. Every day, I had a

Lucid

Russian casserole for either lunch or supper and chose a selection of dehydrated foods for the remaining meal.

We ate all our meals together. Breakfast was generally a fast meal. We ate, then got down to the activities of the day. We had lunch after all of us finished our exercise. Typically, we ate lunch fairly leisurely. We would eat and discuss the activities that we had been involved in. Supper was late in the day, and we generally ate after the major portion of the day's activities had been completed. Many times, we lingered over this meal, laughing and joking together. This was also the meal where we would try novel combinations of the food options. One person would take one of the packets and mix it with another to create a new taste and, if it turned out great, recommend it to the other crew members. A few of the Russian containers had small cans of cheese. These cans were highly prized, and we would heat them up, open them, and share the cheese, mixing it with a packet of mashed potatoes or using it as a dip for a packet of dried fruit.

Russians love mayonnaise, and Yuri and Yuri were certainly no exception. Each American container held a baggy filled with condiment packages: ketchup, mayonnaise, taco sauce, and Tabasco sauce. We used these packets in various combinations to get a different taste with the food. Now, I like mayonnaise, but not with the same fervor as the Russians do. I often traded my mayonnaise packets with Yuri and Yuri for a Russian cookie package or some such delicacy. One evening, I asked them what they thought of Americans living on Mir with the Russians, and they both diplomatically concurred that it was a good thing. I told them they only thought it was a good idea because without the Americans, there would be no mayonnaise! They laughingly agreed with me.

The politeness and thankfulness that Yuri and Yuri exhibited at every single meal caught me by surprise. I raised three children. I provided three meals a day for these children for many years, and I was only too aware that sometimes when a certain food product is repeated too many times, thankfulness runs a little short. But not with Yuri and Yuri. At the end of every single meal, they invariably said, "That was very tasty. Thank you so much." And they always politely excused themselves as they floated away from the table at the conclusion of our meals.

One day we started talking about our favorite foods. Yuri and Yuri told me about their favorite Russian dishes and how their wives prepared them. We began talking about the various restaurants close to JSC that many of the workers frequented for lunch. Both of them had spent time at JSC training as backups for the Russians who launched on STS-71, so they were familiar with these restaurants. Never before had these lunch venues sounded so good to me. Then I carried the conversation into the realm of the gastronomical delights of pizza, hot dogs, and barbecue. I could feel the nibbling of discontent edging around my stomach. I announced that it was time for a new rule. We must never, never talk about any food that we did not have there on Mir. Yuri and Yuri agreed wholeheartedly, and we never again got into a conversation about our favorite foods.

We used the various foods that arrived as gifts for us on the Progress vehicles as a way to make meals and days special. One night, after we finished eating, Yuri Onufriyenko asked what we had special for a treat. He suggested we share one of the cheese-and-cracker packets my family had sent me. I told him that wouldn't work because we had already agreed that we would only eat those on Saturday night because they reminded us of picnics, and Saturday is the day for picnics. Since it wasn't Saturday, he suggested we eat one of the small cakes that we had. I reminded him that we had set those aside as a Friday night treat to celebrate the end of the workweek. Then he asked, "Well, Shannon, what can we have now, on a Thursday night?" When I pointed out to him that we did not yet have a specific way of celebrating Thursday night, he said with resigned disappointment, "Well, so what is a crew to do on a Thursday night?"

One big disappointment with the American food was that there was only one package of shrimp cocktail in each American food container. Shrimp cocktail was the favorite food for each of us. Later, I found out that we had only the one package because it was high in sodium, and the nutritionists wanted to make sure that we did not have too much sodium. So once a week, we hydrated our shrimp cocktail package and shared it with each other. We each got about two shrimp, and we divided up the hot sauce. Not a drop went to waste.

In the Russian food was a type of nut that Yuri Onufriyenko loved. Yuri Usachov and I did not particularly care for them. One day, I found a small white bag from a previous Progress way back in a crevice of the transfer node. It contained packets of dried fruit, which Yuri Usachov and I really liked, and about twenty packages of Yuri Onufriyenko's favorite nuts. I floated into the Base Block when no one

Lucid

else was there and slid the nuts under the bungee cord at Yuri's place at the table. The pleased look on his face when he discovered them was worth more than several rainbows.

One day, I found an American container packed with pudding. There were thirty individual portions. I showed my find to Yuri and Yuri and then explained that we each had ten. I wrote each of our names on the side of the pudding container and told them that each time we ate one, we would put a mark under our name. In this way, we could each eat our share when we wanted to and not worry about eating anyone else's. Of course, this led to bartering. Once, the macaroni-and-cheese packet in one of the American containers went for three chocolate puddings!

The Russian cans had to be opened with a can opener. The one we had on board was the small kind that you might take camping with you. Of course, it tended to get a little messy because cans were opened at every meal. I started to clean it every Sunday night by wiping the accumulated grease and food particles with a damp cloth. Yuri Usachov saw me doing it and laughed. He said that was what he had done on his previous Mir flight.

Anytime I had the passing thought that maybe the food was getting a tad bit monotonous, I had only to say to myself, "Yes, but remember, you didn't have to think up what to prepare for the family for supper, you didn't have to spend Friday night shopping for groceries, and you didn't have to do any dishes—you only crush the cans in your hands, stick them in the outer wrappers for the food packets, and push them into the trash. What could be simpler?"

FORM 24—OTHERWISE KNOWN AS COMMANDS FOR THE DAY

Each day's activities were detailed for us in a daily timeline, which the Russians called a Form 24. The Form 24 for the next day was uplinked to us every evening. After it was printed out in abbreviated Cyrillic on a narrow strip of paper, we would pass it around so each of us could highlight our individual tasks for the next day. For the cosmonauts, these tasks were typically Mir maintenance activities, whereas I would be timelined to conduct the various NASA experiments that were on board. In my notebook, which I kept stored in the top part of my coverall (and yes, *coverall*, not *coveralls*—I wore the same one every day for six months). I jotted down what I was to do and when.

We, the crew, usually looked at this detailed timeline as "permission" to do certain activities the following day but did not adhere to the times given for the various tasks. We did the scheduled activities at the time that it made sense for us to do them from the onboard perspective. At the end of each day, Yuri Onufriyenko would hold that day's Form 24 and run his index finger down it and ask Yuri and me if we had completed our activities for that day. If we had not been able to complete a scheduled task, then we talked about why it was not completed.

Lucid

Exercise

Generally, regardless of when the Form 24 scheduled us to do it, we all exercised just before lunch. Exercising all at the same time was the most efficient use of our time because in this way we did not interfere with another person's work. Aboard Mir were two treadmills. The one in the Base Block was located, incidentally, right next to the table and right in front of the hatch that led into Kvant 1. (If you had tried, you could not have found a more inconvenient place, a place more disruptive of anyone else getting anything done, than this location.) The other treadmill was in Kristall. Although more convenient, this treadmill was not totally functional. In the Base Block was also the ergometer. This was stored under a floor panel and only taken out when being used.

Daily, we rotated the exercise equipment that we used. Over the years, the Russian exercise physiologists had developed protocols that involved running on the treadmill and using bungee cords for resistance training. There were three different protocols, and crew members used a different one each day. Generally, a protocol could be completed in forty-five minutes. Toward the end of my time on Mir, I felt that I needed to be working harder, so after I finished the written protocol, I ran additional kilometers on the treadmill.

One day I was running as fast as I could on the treadmill in Kristall while Yuri and Yuri were busy in the airlock checking out their space suits for their upcoming spacewalk. I was lost in the musical oblivion of my headset. Suddenly, Yuri's face was floating one inch in front of mine, mouth going ninety miles a minute. I did a startled float off the treadmill and whipped off my headset. Above the racket of the still-running treadmill, I heard Yuri's words, "Stop! Stop your running, Shannon! You are tearing up Mir!" And then he explained to me how back in the airlock, my running was causing the airlock to oscillate around them and the space suits. They were afraid that these periodic oscillations my running had induced on Mir would cause some type of structural failure. From then on, we never used the automatic mode on that treadmill.

I will be honest—the daily exercise was what I disliked most about living on Mir. For me, the most negative aspect of the exercise regimen was that it was just downright boring. The treadmill was so noisy that you could not carry on a conversation with another crew member. Also, it was hard to maintain any level of excitement or interest when

Tumbleweed

doing the very same thing every third day for six months. I listened to my Walkman while running, but I rapidly realized that I had made a huge preflight mistake. I had packed very few tapes with a beat fast enough to run to! (When I heard that my stay on Mir would be extended, my very first thought was, "Oh no, one and a half more months of daily treadmill running! Seven more Saturdays of ritual running and listening to my Carman tape, *Sunday's on the Way*.) On Mir, there was a large collection of music tapes that had accumulated over the years. Believe me, it was a red-letter day when I stumbled across an Elvis tape—no, not because it was Elvis, but because it was in English and not Russian! During my six-month stay, I worked through the majority of the Mir tape collection.

Three things kept me motivated, day after day, to exercise. The first was that while I was exercising every day, I thought about returning home, reminding myself that I wanted a body worth coming home to! And every day while I was running on the treadmill, I would replay an imaginary video in my mind of the shuttle landing, the hatch opening, and me walking out into the sunshine. The second was a little negative motivation, in that the receptors in the bottom of my feet became very sensitive if I did not run daily; if I skipped a day without pressure on the bottom of my feet, it was very uncomfortable to start running again. My feet would feel prickly, somewhat like the pins-and-needles feeling you have when standing after the blood supply has been restricted to your feet. The third reason was that it felt so good when it was time to stop.

I could get ready to exercise in five minutes flat. I would pull on my pair of shorts and T-shirt that I had brought over from the shuttle—yes, I wore the same T-shirt and shorts every day for six months. The reason was simple. When I put on the Russian shirt and shorts set, which was made out of thin, inexpensive T-shirt material (read, *see-through*), I looked down at myself and thought, "Oh my goodness. I can't let anyone see me like this! After all, I am a grown woman and not some cute teenager!" The shorts did not hang around my legs like shorts should; they curled up into the fold of my crotch. Sure, one size fits all, but there are fits, and then there are fits. The fit of the Russian shorts set on me was definitely not for public consumption!

I just about panicked, thinking, "I cannot go out in public like this—I have to exercise. What can I do?" And then I jumped to the only possible conclusion, and that was that, obviously, I would have to

Lucid

wear my only pair of shuttle shorts and T-shirt every day. "OK. It is no problem, Shannon," I told myself. "It is just for one hour every day—they can dry out overnight." And dry out they needed to, because many times, the area where we were exercising was above ninety-five degrees, with humidity higher than in Houston, Texas. The sweat would literally pool all over your body—it didn't run off in microgravity; viscosity maintained it as a sheet of water covering your entire body. Soon, the shuttle shirt and shorts set lost the soft cloth feeling and became literally stiff like a board with the accumulated dry sweat.

Every day, after I floated into my stiff shorts and shirt, I slipped into my harness—sometimes still wet from the previous day—and then put on socks and tennis shoes. Wait—tennis shoes? Why shoes, you ask? You don't use your feet in microgravity, do you? True, except when you are running on the treadmill. And there you really need them! I know; I tried to run without them. I agree, stupid, but desperate! One day, I could not find one of my shoes. I looked everywhere—behind this panel and under that panel. I knew it had to be on Mir. I'd had it yesterday. No hatch had been opened. No extravehicular activity (EVA) had been done, so the shoe had to be physically present on Mir. The only question was where. Not wanting a little problem like a lost shoe to stop me, I decided to just run shoeless and promptly got on the treadmill barefooted.

What a mistake! After a few minutes, my feet were killing me. A treadmill is a continuous sheet of heavy rubberlike material placed over rollers that are turned by a motor, which in turn causes the movement of the track. Running barefoot meant that each of these rollers impacted my foot through the rubberlike material. It felt like sledgehammers hitting the bottom of my feet. I quickly gave up on trying to run. And immediately, visions of myself returning to Earth as a floppy jellyfish started to dance in my head.

That night at supper, I told Yuri and Yuri about my lost shoe and explained that I was desperate. I had hunted everywhere, to no avail. I needed my shoe. I needed their help. Please would they help me look for it? Of course, they agreed, but they needed some motivation. Something to whet their hunting instincts. They needed a prize. "OK, OK," I said. "The person who finds my shoe will get a surprise." They looked at me.

Tumbleweed

"What kind of surprise are you offering?" they asked. After all, they knew what was on Mir. They knew every nook and cranny even better than I did, and nothing jumped immediately to mind as a suitable prize! So again, the question, "What can you offer us that will be worth the effort expended on this hunt?"

I thought a moment and then said, "A bag of Jell-O all to yourself, a bag of Jell-O that you do not have to share." And they immediately started tearing Mir apart, searching for my lost, floating shoe.

At this point, it may legitimately be asked, why did this Jell-O offer such a gigantic incentive? Publishers Clearing House knocking at the hatch could not have elicited such frenzied activity—after all, there is nowhere to spend a million dollars on Mir! This was better than if they had been told there was a golden egg hidden somewhere—more about hidden eggs later—because after all, on Mir, you could not really do anything with a golden egg!

But that Jell-O! To this day, I can vividly remember how we savored each small spoonful we had. During training for the flight to Mir, I learned that there was a refrigerator located in the Base Block. This opened up new possibilities as far as food was concerned because on shuttle flights, we did not have access to a refrigerator for food products. And that was OK because shuttle flights were so short. I tried to think of what simple food product could be brought along to utilize the Mir refrigerator. The logical conclusion was Jell-O. Before flight, it could be sealed into the already-approved drink bags. Then, in orbit, all the crew would need to do would be to add hot water (just like was done for tea or coffee), shake the bag, and place it in the refrigerator until it gelled. Then the top could be cut off, and the Jell-O could be eaten using a spoon.

The food group at JSC was very happy to do this for me, and I still remember the first time I surprised Yuri and Yuri with a bag of Jell-O. Their delight with the unexpected treat was gratifying. After that first offering, I counted the bags we had on board and determined that we had just enough to last through the entire time the three of us would be on Mir together if we had only one bag every week. We decided that sharing a bag of Jell-O would be our Sunday night tradition. Every Sunday I prepared a bag, and then, in the evening, after the last comm with mission control, I cut open the bag and we passed it around—a spoonful for you, a spoonful for you, and a spoonful for me.

Lucid

That evening, after I had made my offer of a Jell-O prize, Yuri and Yuri literally tore Mir apart. Panels were floating everywhere. Suddenly, Yuri Usachov shouted the Russian equivalent of "eureka!" He then drifted into the Base Block with my prodigal shoe. The other Yuri and I watched as he ate his prize. He offered to share, but we insisted that he eat it all—after all, he was the winner! Having a pair of shoes to use while exercising was certainly worth a bag of Jell-O!

One of the big motivators for me to exercise was to slow down the loss of muscle mass in my body. There was no doubt that our bodies were changing. We could see the change in our own bodies and in each other. The first Progress that arrived brought mail from home, and in that mail we found a few pictures that the STS-76 crew had taken of us. In these pictures, our faces still looked a little puffy. This is the natural look when people start living in the absence of gravity. Without gravity to pull body fluids downward, the fluids become redistributed in the body, and this redistribution fills the tissue in the face, making it appear a little bit, or sometimes a lot, puffy. Over a period of several weeks, the body gradually gets rid of what it perceives as excess water, and the puffiness disappears. We passed around the photos and compared the way we looked then with the way we currently looked. Each of us had cheekbones that were more accentuated. After several months of not having gravity tug our cheeks downward, each of our faces had a more triangular look. We agreed with each other that we definitely looked at least ten years younger than when we launched!

The new shape of my face was not the only change that I noticed was happening to my body. Every day when I switched shirts to exercise, I thought that I was in a snowstorm! The act of pulling my shirt over my head released a cloud of tiny dry skin particles that floated around me like flurries of snow. I had never realized that one purpose of taking a shower was to wash off all the particles of dead skin. Sometimes in the late evening, all this dry skin would cause my arms and legs to itch. I looked in the well-stocked medical kit I had on board to see if there was anything there for dry and itchy skin but found nothing, so I improvised. I discovered the perfect way to handle this problem. I took a piece of male Velcro and rubbed it over my arms and legs. It just felt so good to rub off all that old skin.

But it was the changes in my feet that amazed me the most. Because I was not using my feet, the years of accumulated calluses on my soles began to come off. Sometimes in the evening, I could pull off big

Tumbleweed

chunks of this dead skin. Before I returned to Earth, the bottoms of my feet were as smooth and soft as that of a newborn baby. All my life, I have always gone barefoot. The second evening that I was home, I naturally ran outside barefoot to pick up the newspaper. As the sharp, dry, dead pine needles in the front yard pierced my baby-soft feet, I realized why feet form calluses!

One evening after about four months in orbit, while we were all floating in the Base Block waiting to establish comm with the ground, I took off my socks (I generally wore socks to keep my feet warm) and began examining my feet. To my surprise, I noticed that I had a very tough callus on the top of my left big toe. There was no callus on the top of my right big toe. Then I realized that I always hooked the big toe on my left foot under one of the conveniently placed metal foot loops to stabilize myself whenever I wanted to remain fixed in one place. Before I saw this callus, I had not realized that I always used the big toe on my left foot for this purpose. I asked Yuri and Yuri to look at their feet and was surprised to find out that they had calluses on both of their big toes. Obviously, we had been fixating ourselves differently. They must have been using their feet interchangeably, whereas I had only been using my left big toe.

That evening, I sent an email note to my daughter, the aspiring writer, and told her that I had a great plotline for her if she wanted to write a mystery novel. The plot would revolve around the discovery that the murder victim had a callus on the top of his big toe, which proved that he had been in space at the time of the incident—you get the point. Unfortunately, she has not yet seen the brilliance of this novel plotline, and the mystery has yet to be written.

After each exercise period, it was cleanup time. Invariably, I am asked how we took showers on Mir. The answer is simple: we didn't. Although I did not take a shower for six months, I managed to stay clean. I know this because one of the first questions that I asked my friends on STS-79 when they arrived on Mir was, "Now, we are friends, so I want a truthful answer. Do I smell?" And the answer was no. There was my proof. I was clean—at least I did not smell!

Keeping clean was simple. From the shuttle, I had brought over sealed bags that contained a no-rinse soap solution, which were used on shuttle flights for sponge baths. When these bags, which were just like the drink bags, were filled with hot water and a straw was inserted, a controlled amount of hot water could be cupped in my hand.

Lucid

Viscosity held it together as a ball in my hand. Then I would lift one side of my shirt and gently rub this water over half of my body and repeat for the other side. Then I would towel dry under my shirt. After I was clean and had "rinsed" the sweat from my body, I would remove the exercise shirt and put on the Russian T-shirt that was the current one for that week.

Once a week, just before bed, I would wash up more extensively. First, I'd locate myself in Spectra with the lights off. It was pitch-black. Then I would remove my shirt, apply warm water, and rub myself down very hard in the dark. I kept the clean shirt in my teeth so that if the master alarm went off while I had my shirt off, I could whip the clean one on while floating posthaste into the Base Block to determine the nature of the emergency. After I cleaned the top half of my body in this way, I repeated the process for the bottom half.

After my "bath," I would wash my hair. To accomplish that, I would use no-rinse shampoo. I had both American and Russian versions. I could not tell any difference in how each worked, but I alternated between them just in case there was something different. I figured that my hair deserved a little variety. To use the shampoo, I would just gently squeeze some into my hair and then, just as gently, squeeze some water from a straw into my hair. Then I would ever so gently mix it and rub it into my hair and scalp until my entire head was wet. Then I would take my towel and vigorously rub my head and hair to soak up all the excess liquid. Now, don't get the idea that it was a clean towel I was using! I used a clean towel once a week, letting it dry after each use. I used the towel for my head that I had used for my body the previous week. After my weekly bath and weekly change of clothes, I would feel clean and refreshed, and it felt good that evening to snuggle down into my sleeping bag, smelling my clean hair.

EARTH OBSERVATIONS

One of my favorite places in Mir was the large round window located in the aft part of Kvant 2. This window was out of the normal traffic pattern of our daily living on Mir. Actually, the only reason to be back there was to look out the window or to work on space suits located in the adjoining airlock. At least once a day, I would float back there, turn the crank to swing open the cover that protected the outside of the window, and hover with my nose pressed to the glass. Generally, the position of Mir was such that during the daylight portion of the orbit, this window looked straight down at Earth, only a twenty-day walk away in physical distance. In accessible distance, though, it rested at the outer limits of technological capability. Without fail, the first glance always caused me to catch my breath and involuntarily exclaim the words from Psalm 121:1–2, "I will lift up my eyes unto the hills from whence cometh my help. My help cometh from the LORD who created the heavens and the earth."

During the night portion of our orbit, this window generally pointed out into deep space. Invariably, the seemingly infinite array of stars flung out on the infinite black-velvet universe made me think of God taking Abraham by the hand and showing him this scene as they covenanted together. Then I would recite the poem, written by my daughter Kawai, that I had found tucked into one of my books waiting for me on Mir:

Lucid

In Space

In space, can you see the sunrise?
Do roosters crow the dawn?
Does sunlight pierce the darkness
proclaiming night is gone?

In space, can you smell creation?
Is the breath of heaven new?
Does the fragrance of burning starlight
smell fresh as morning dew?

In space, can you hear the footsteps
of God upon His floor?
Do you listen to songs of angels?
Do you cry aloud for more?

In space, can you taste a moonbeam?
Will a sunray burn your lips?
If you taste now of forever,
will you be content to sip?

In space, can you touch tomorrow?
Is the future night or day?
Does eternity stretch before you
and beckon you to stay?

In space, do you know we miss you
and wish that you were here?
Does memory say we love you
and always hold you dear?

Over the course of months of daily Earth-viewing sessions, I realized that one of the great things about a long spaceflight, as opposed to a short shuttle jaunt, was the fact that I could appreciate the flow of seasons across the face of our planet. I arrived on Mir at the end of March. The first view I had of the Northern Hemisphere was ice, snow, and more ice. All the lakes were frozen, and the ground

Tumbleweed

was white with snow. Within a few weeks, huge cracks could be seen in the lakes as the ice started to break up. Seemingly overnight, the Northern Hemisphere glowed green with spring.

In addition to the seasonal changes, there were the dramatic single-event occurrences on Earth. One day as we passed over Mongolia, we saw huge plumes of smoke. It seemed as though the entire country was engulfed in these smoke plumes. This amount of smoke seemed unusual enough to mention it to the ground. Days later, they informed us that word of huge forest fires was just starting to filter out of Mongolia and appear in some newspapers.

Using the onboard binoculars, I saw the icebergs leaving Antarctica. A few orbits later, on one of the few occasions when it was not covered in clouds, I saw Greenland and what appeared to be thousands of icebergs being born from its glaciers. One time, the same orbit, I saw both northern and southern lights, green wisps of witch's hair streaming through the atmosphere of Earth.

I quickly noticed that we spent far more time passing over water than land. From space, it is obvious that our Earth is truly a ball of blue water with white clouds swirling through it and only a few blobs of landmasses floating here and there. As we passed time and again over the landmasses, I was struck by how small, in a relative sense, the United States is. We would go from the west coast of America to the east in about ten minutes, whereas it would take over forty to transverse the Eurasian landmass. Every time we would orbit it, I would think, "So many countries, so many people that I've never had the opportunity to encounter."

The great thing about this time spent observing our Earth was that I was actually working while doing it, not just being a space tourist. Many different investigators had requested photographs of various portions of our planet under varying seasonal changes and lighting conditions. All these photographs are now being used to better understand what is happening on Earth.

A favorite topic of conversation at mealtimes was our daytime observations through the windows. Yuri Usachov would invariably add a few comments about the silver clouds he saw on his first flight. He described them as slim silver streaks above the thin shell of our planet's atmosphere. Once, after we had said our good nights, he flew into Spectra and shouted, "Come, Shannon. I see silver clouds!" I flew after him to a window, and there, sure enough, was a silver streak in

Lucid

the blackness of space, high above the thin blue line of Earth's atmosphere. We saw this strange phenomenon twice more during my six months.

Strange as this observation was, it did not compare to what I saw on my first shuttle flight, STS-51-G. It was the end of our workday. I was alone on the flight deck. The rest of the crew was on the middeck, involved in their evening preparations. I was alone at the back windows, staring out into the vast blackness of space. Suddenly, a brilliant orange ball appeared, rising from Earth. I gasped, certain I was seeing an atomic blast. After all, this was in the 1980s, and the Cold War was still going strong! Then I realized it was a ball. Not moving and not taking my eyes away from the window, I called down to the middeck for someone to come up and look out the window. Suddenly, the orange ball fragmented into a myriad of orange shards—to me, looking all the world like a burst balloon. For an instant, it was dark. John Fabian floated up to my side, and I pointed out the window just as nine orange triangles, in three groups of three each, popped up where the "burst balloon" had been. The triangles appeared to be headed earthward. Before John and I could verbalize what we were seeing, the orange triangles extinguished, and there was nothing but blackness out the window. John and I looked at each other in stunned silence. What had we seen? Whispering to each other, we concluded that maybe we had witnessed a rocket launch of some sort. We both agreed that because we knew nothing about it, we should not say anything to the ground. We got Dan Brandenstein, the commander, up to the flight deck and described what we saw, and he agreed with us that we should not talk about it until our debrief after the flight was completed.

After landing, Dan debriefed our boss on what we had seen and was told to tell us to never talk about it to anyone. No reason was given for the need for silence. For years, I just wondered what we might have seen, and later, I surfed the internet for clues. John and I agreed that we must have seen some type of missile launch. The "bursting balloon" could have been the nose cone exploding. The nine separate orange triangles that appeared immediately after this could be explained by nine separate "vehicles" being released and then simultaneously igniting their engines. Years after being told not to mention this observation, NASA management changed, and I explained what I had seen and asked if it would be all right to talk about

Tumbleweed

it. "Sure, if you want folks to think you are crazy" was the reply. Being thought of as crazy certainly did not bother me, but I never seemed to be involved in a conversation where decades-old space observations seemed apropos.

One morning, I took a ten-minute "observation break" from work to check up on what was happening in the United States. That day, the entire country was cloud-free. As I watched first the Cascades in Washington, then the Rocky Mountains, the Colorado Plateau, and the mighty Mississippi pass beneath me, I began an off-key rendition of "America the Beautiful." Just as I got to the waving grain, Yuri Usachov, who had soundlessly floated to my side, asked me what I was singing. Laughingly, I explained. He asked me to teach him the song in English so that we could sing it every time we went over America. We worked on it sporadically at suppertime, and for the rest of the flight, whenever we were together in the observation window in Kvant 2 as North America rolled beneath us, we would sing, off-key, "O beautiful for spacious skies." That was as far as our duet ever progressed.

EVENING

My favorite time of the day came right after the last comm pass, usually between 10:30 and 11:00 p.m. Right after we said our good nights to the Moscow control center, Yuri Onufriyenko would invariably ask, "Now is a good time for what?"

Yuri Usachov and I would laughingly reply in unison, "It is a good time to say good night!"

While saying my good nights, I would be concurrently filling two drink bags and refilling, for the tenth time, my bag that contained a tea bag. Tea was a precious commodity on Mir, so I used the same tea bag over and over. My filling of the drink bags was always accompanied by the comment, from either Yuri or Yuri, "I do not understand how Americans can possibly drink so much water!" After the completion of this nightly ritual, I would say a final good night while floating off into Specter, loaded down with my drink bags.

Arriving in Specter, I would untie the string that kept my sleeping bag rolled up during the day and then tie it to a handrail that was fastened to the floor. In less than ten seconds, my bed was ready. In the morning, the process was reversed. I loved the elegant simplicity of preparing my sleeping place. In fact, it could be said that in some ways, I had been training for this all my life. As a child, I never could understand my mother's fixation with making beds. "I have too much

Tumbleweed

life to live and cannot waste precious moments every morning making my bed" was my constant wail to her. I tried various subversive techniques, from sleeping on top of the spread to sleeping in chairs to sleeping on the floor, all so I would not have to make my bed in the morning. But nothing worked. My mother decreed that I should sleep in a bed with sheets. So every evening, I always had to unmake the bed only to make it again come morning. So, you can see why I loved the elegant simplicity of preparing and unpreparing my sleeping place on Mir.

As soon as my sleeping bag was positioned, I settled down into my nightly routine. If I'd had the good fortune to receive some email that day, I would get out the computer and read it. Received email or not, I would type out a message to my husband and children, copy it to a floppy, and put the floppy in the zippered pocket on the leg of my coverall. I was now prepared to downlink my messages the next day if a packet link was established via the ham radio. That done, I would get out my small Bible and read "the Psalm for the day."

The idea of reading a Psalm a day came to me during church the last Sunday I was in Moscow. During the service, someone read the 150th Psalm. When the reference to Psalm 150 was intoned, a light bulb flashed on in my skull. There were 150 Psalms, and according to the plan at that time, I was to be on Mir 144 days. Starting from the first day that I was in orbit, my family and I could read one Psalm every day. Prelaunch, having no idea of the "whens, whats, or wheres" of how we would be able to keep in contact, this seemed like an ideal way of knowing that we could at least share one common thing each day. Since I was to be on Mir only 144 days, there would even be a few Psalms left over to take care of any contingencies that might happen to keep me in orbit longer. Little did we know!

Every evening, I read my daily Psalm and thought about my family reading the same one for that day. Then I would choose a verse from that Psalm that meant the most to me right then, copy it out on a yellow sticky note, and stick it on a panel to read and think about the next day. (This was OK. No one else could read English. There were times when Spectra appeared to be wallpapered with sticky notes!)

The Bible that I had with me was the small one that I had gotten many years before when I was in the fifth grade. During my childhood, books were very expensive and hard to come by. (I was a child before paperback books were ubiquitous.) To me, it seemed that my entire

Lucid

childhood was one eternal quest to find books to read. There was a bookstore close to where we bought groceries, and every week I would go in there to hungrily gaze at the books. A small Bible caught my eye. I wanted it more than just about anything, but it was almost twenty dollars. Such a sum I would never be able to save up. I talked my parents into agreeing to give it to me as a reward for reading the entire Bible through, from beginning to end. After a year and a half of daily Bible reading, I finally got to the "I am the Alpha and Omega," and the coveted Bible was mine. Ever since that day, it has traveled with me everywhere I have gone.

After time with the Psalms, it was time to turn my attention to the current book I was reading, although ensuring that I would have enough books to read on Mir was a more difficult task than I had imagined.

Soon after Norm Thagard returned from Mir, the NASA personnel listening to his debriefs realized the importance of some form of personal recreation on a long-duration spaceflight. Late one evening, after an all-day payload training session in Houston, two people who had been given the task of providing psychological support for the Americans on Mir came up to me and asked what I thought I needed in the way of recreational activities to be happy on Mir. My answer was simple: "Just make sure I have plenty of books, and I will be happy."

"No," they said. "You don't understand. NASA is committed to providing you with recreational outlets. Maybe movies?"

"No, thanks. I don't watch movies often now, so why would I watch them on Mir? Just some books, please."

"Well, maybe you would like to take a short course in your free time?"

"No, thanks, just books. If I don't run out of reading material, I will be happy."

I went back to Russia, and there I got a message from my family that NASA had called them and said that some books could be sent to Mir for me. There was room on STS-74, the shuttle flight to Mir that would bring up the docking module. But the books had to be turned in within ten days.

The good news was that NASA listened to me, and I had lots of books on Mir. The bad news was that because I was in Russia for training, the book selections had been at the mercy of my daughters, one of whom had majored in English in college and the other of whom

Tumbleweed

would never spend money, even her dad's money, on a new book when you could find a perfectly good one at Half Price Books.

The first book I read on Mir was *The Pilgrim's Progress*. It was a book I had always meant to read. Previously, I had read only an abridged version when I was in the third grade. After *The Pilgrim's Progress*, I progressed to Dickens. For some reason, I had never read *David Copperfield* and *Bleak House*, so they were next. My daughters had thrown in some books I had left piled by the side of my desk that I had been planning to read but had not yet gotten around to. That was the reason why I had the complete works of Flannery O'Connor on board. I read one short story a night until I had read them all. One story at a time was all I could take! And then it was back, courtesy of my daughters, to seventeenth- and eighteenth-century England! Don't get me wrong, I like English literature, but after a few months I was really longing for anything that would be just a little change of pace!

As I would start yet another book by an English author, I would remember the conversation my daughter and I had when I asked her to select my Mir library, which was impossible for me to do from Star City.

"Please find some books for me and give them to NASA. Fast!"

"What kind of books? I certainly don't know what you have and haven't read and what you might want to read in outer space."

"It doesn't matter what kind. Choose books that are word dense. Books with the most words per page. I don't want to run out of words!"

"OK. I'll choose some that I know you haven't read but that you should have read. Books that will be good for you."

This should have been a warning to me; after all, I did know my daughter was an English major in college. "OK, OK. Just remember—lots of words on every page." And that was how *Tom Jones*, maybe the longest novel in the English language, ended up on Mir. Fortunately, my stay was not extended to the point that I was forced to read it!

I read several books straight in row: *The Mill on the Floss*, *The House of Mirth*, *Middlemarch*. Each of these books had one thing in common—the author did not know what to do with a strong female protagonist except to kill her off on the last page. I felt like shouting across the ages to them and saying, "Don't you know, females can live in this world and do useful work! They can really live! They don't have to be killed off just because they want something more than what

Lucid

society would allow them to have!" I would mumble, grumble, and rant to myself.

Almost every night as I was reading, I would be cognizant of where I was, one of only three human beings in the entire universe not living on the planet Earth, and then I would think about the authors who had written the books that I was reading. Never in their wildest imaginations did they ever dream that their words, some being penned by candlelight, would one day be located in a library on humanity's cosmic outpost, bringing tears or peals of laughter to one of the three human beings locked in a tin can hurtling around Earth at 17,000 mph, savoring their every word. And every night, I was humbled by the power of the written word.

Living in $0\ g$ is fun, but like everything else in life, it has its pluses and its minuses. The physical act of reading books printed on paper in $0\ g$ is definitely on the list of minuses. My favorite way to read a book here on Earth is to scrunch down into a big recliner and just wiggle until I can't tell where I end and the recliner begins. The book rests on my pulled-up knees; I do not have to hold it. It is entirely different in $0\ g$. There, you don't have gravity pulling the book open. You must physically hold the book open, and believe it or not, that gets a little tiring. And then you have to anchor yourself by sticking your toe under a handrail. This is not the comfortable, scrunched-up way to read. I always had to put forth an effort.

One day, I made myself a "space chair." I took some big pieces of foam that had been used for packing science equipment for launch and now was just so much trash that I was cutting up to fit into the Progress. I took a piece, rolled it into a cylinder, taped it so that it would not open, and then slipped it under a bungee cord to hold it in place on the Spectra wall just where my neck would rest. In a like manner, I strategically placed a foam roll where the small of my back contacted the wall, and there it was, my "space chair"! By bending my legs and putting my feet on the opposite wall, I could almost get into my most favorite scrunched-up reading position. It worked great, but I could never figure out a way to keep the book open without effort.

One day, Yuri was working in the back end of Spectra, trying to impose a little order on the jumble of unused science equipment floating there, and he started to take down my foam. I hollered, "Wait! That's my chair! Don't take it away!"

Tumbleweed

He shook his head and gave me his "Americans sure are weird" look, shrugged his shoulders, and said, "Whatever makes you happy!" It stayed in place for the duration of the flight.

At twelve o'clock sharp every night, I turned out the lights and floated into my sleeping bag. The good nights were those when Spectra was nice and cold. I could then snuggle way down into the bag with my feet wrapped in a sweater, which I kept in the foot of the sleeping bag. A couple of times I tried to listen to my Walkman to go to sleep, but I fell asleep so fast that I did not get it turned off. Because batteries were a limited commodity, I gave up on the idea of going to sleep to the sound of music.

Sometimes just before going to sleep, I would see flashes of orange light on the back part of my eyeball. I realized that this light show was caused by cosmic rays hitting the retina of my eye. I had observed this only one time before, on my first shuttle flight. The orange flashes were of two distinct types. One type was a streak that was a narrow band of bright light. The other was like a soft baby light bulb switching on and off. But most nights, I fell asleep faster than my eyes would adapt to the dark, so it was the rare evening that I experienced this phenomenon.

All my life, I have always looked forward to nighttime because it is once again the time of day when I can go to sleep and dream. I love to dream. I love the swirling, vivid dreamworld of colors, brighter than real life; the exciting action; the cascading emotions; and the brilliant story lines, which, unfortunately, rarely retain their brilliance in sunlight. I found that being in space did not change my pleasure in dreaming. The colors, the plotlines, and the excitement remained. But one thing changed. When earthbound, I have never dreamed that I was floating in space.

Now, I have dreamed that I was *in* space—working in the space shuttle, for instance—but in all those dreams, neither I nor any of the other dream people were floating. Imagine my surprise when I found that after I had been in space for a few weeks, my dreams were populated with floating characters. The setting was Earth, but Earth minus gravity. For instance, one very vivid dream took place in Clear Lake City, where I live. I was inside my blue Datsun with three other people. We were stopped at Highway 3 and NASA Road 1. We were being held up by bandits. I was in the back seat, and then I *floated* into the front seat to disarm the bandits. Weird!

Lucid

The only other thing I noticed that was different from my normal dream pattern was how intense the dreams were after a few months on Mir. Sometimes I would wake up in the morning with a feeling of relief because I needed to rest after all the intense emotional activity of the dreams. My personal theory is that this was a result of a lack of stimulus during the day.

All too soon, every evening, the alarm I had set for midnight went off. I was very disciplined about going to sleep on time because a good night's sleep is one of the three things that I found necessary to be happy in space. (The other two things are eating well and keeping well hydrated.) No matter what I was doing, I always turned off the lights and floated into my sleeping bag promptly at midnight because all too soon, the 8:00 a.m. alarm set by the Moscow control center would ring throughout Mir, announcing the start of another day.

PROGRESSES

Every outpost must have a way to get supplies. Mir was no exception. It was resupplied by Russian Progress vehicles. The Progress vehicle was the same as the Soyuz, which was used to ferry crews to and from Mir, minus the life-support system. In other words, instead of crew members and all the auxiliary equipment needed to maintain them on a voyage to Mir, a Progress vehicle was loaded with the supplies that the team on the station needed. After launch, it automatically flew up to and docked with Mir. After all the new supplies were unloaded, the now-empty Progress was filled with all the trash that had accumulated on Mir. Once full of trash, it was undocked and burned up upon reentering Earth's atmosphere. A system in which the supply vehicle doubled as a garbage truck was a functionally elegant system indeed!

Progress I

Although the shuttle that brought me to Mir also delivered about two tons of supplies, some supplies were beginning to run low after I had been there for about two months. We eagerly anticipated the first of two Progress vehicles that would arrive on Mir during my stay. It was as if I were a kid counting down the days until Christmas. Of course, we were looking forward to having our supplies replenished, but we were most excited about the letters and small packages from home sure to be on board.

Lucid

The Russians have always supported their long-duration flight crews with a ground-based psychological support team. One function of this team was to put together small surprise packages for the crew. These packages arrived on Mir by means of the Progress vehicles. Preflight, I had been asked what I would like in my packets. My answer had been letters from home and reading material. I could not imagine missing anything else. Just shows how little I knew!

While I was still training in Russia, the NASA psychological support team called my family and asked them to provide a few trinkets that I might enjoy on Mir. Later, my daughters told me how they had spent a portion of their precious Christmas vacation wandering around the mall looking for something that might be appropriate to send to a "mother who would be lost in space." (These were their words, not mine!) As they wandered, they commiserated together, lamenting the difficulty of their task. Kawai commented to Shani that they were unquestionably the only people on the planet looking for gifts to send to a mother in space.

As with the Priroda docking, we prepared for the arrival of each Progress by setting up the TORU system in the Base Block. Yuri Onufriyenko would use the TORU system for a manual docking if the automatic docking system did not work properly. And then we waited.

The evening before the Progress docked, Yuri Usachov told me that every Progress was given the name of a horse by the Mir crew. He thought that maybe it would be appropriate for us to choose an American horse for our Progress and asked me to name a few. I thought of Black Beauty and Silver. I personally favored Ichabod Crane's horse, Gunpowder, and tried to relate the story of the Headless Horseman to him, but I guess something was lost in my Russian version because he went with the name Pegasus for our Progress.

Both times, we first saw the Progresses from the window in Kvant 2. Unlike Priroda, the Progresses were not powered by internal batteries but by deployed solar arrays that converted sunlight into electricity. Because these arrays reflected sunlight, they made the Progresses possible to spot at a farther distance than Priroda. As we watched, a spectacular thunderstorm crossed the Atlantic Ocean near the northern coast of North America. There was a dramatic visual display of lightning, rhythmically jumping from cloud top to cloud top, like so many visual tom-toms. Suddenly, above the cities strung out along the coast like Christmas tree lights, there was the Progress, a bright morning star skimming along the top! Soon it was close enough to see the deployed solar arrays, the arms and wings of an alien insect headed straight toward us. In that moment, I really did feel like I was in a "cosmic outpost" awaiting supplies.

After the uneventful automatic docking of the Pegasus, Yuri Usachov released a small valve to equalize the pressure between Mir and the Progress. He stuck his nose close to the outrushing air and said, "Ah, just smell those fresh fruits!" I did not make any comment, but personally, I thought it smelled like opening a refrigerator door that first day back from vacation and suddenly realizing that you had forgotten to empty it before you left! Yuri and Yuri quickly opened the hatch and secured it with a string of locks encircling its entire perimeter. Yuri Usachov slithered over the open hatch, and then half of his body disappeared into the crammed Progress. With a huge grin, he reappeared, waving a white bag above his head. "Lunchtime!" he yelled. We gathered around the table in the Base Block, and he sliced the retrieved fresh tomatoes and onions. We carefully squirted a few drops of our salt solution onto the tomatoes. As we shared this bountiful lunch of fresh produce together, I thought to myself that lunchtime had never been, and could never get, better than this!

After lunch, it was time to unpack the Progress. The supplies had not been loaded with much rhyme or reason. As Yuri and Yuri were pulling out packets of clothes, little packs of screws, towels, or random greenhouse parts, I organized the floating objects into big bags: a bag for towels, a bag for clothes, a bag for "parts." When our personal packages were uncovered, all unloading stopped, and we gathered in the Base Block to go through them together. We were in no hurry. We spent our time going through the packets of letters and clipped magazine articles. We shared our earthly treasures with each other. Yuri showed us the pictures drawn by his kids. I showed them the photo of Michael working with the wolves. Yuri and Yuri shared their perfume-scented letters from their wives.

Although I had not known before flight the importance of edible treats in the Progress packets, there was a huge package of M&Ms in that first Progress. The M&Ms were there due only to the quick reaction time of the psychological support group in Houston. I had done an interview with a radio program and was asked what I missed most. Without thinking, I blurted out that I really missed M&Ms because we had eaten the last ones I had brought over from the shuttle

Lucid

only that week. Each time I searched Mir for any equipment necessary for my work, I kept my eyes peeled for any stray package of M&Ms to appear miraculously. None had, so their absence was uppermost in my mind. Kelly Curtis, a member of the psychological support team, heard that comment and rushed to the drugstore. She purchased a huge bag of M&Ms, found the contingent of NASA personnel leaving for Russia that day, and persuaded one of them to put it into their already overloaded bags. Once in Russia, the M&Ms were handed over to Bill and Gaylen in the Russian control center. They had them hand-carried by a sympathetic Russian on the last plane out to Baikonur before the Progress launch.

The "missing M&Ms" comment struck a chord with listening space fans on planet Earth. On my return from space, the first thing I was handed on the tarmac at the Kennedy Space Center landing strip was an official box of M&Ms from the White House. Don't worry—they were normal, off-the-shelf M&Ms but were "official" because they were packaged in a box labeled with the White House seal. This is the one thing that is remembered most by the public about my Mir flight, and many times when I went somewhere to talk about the mission, I was handed a package of M&Ms. Observing this phenomenon, my son often lamented to me, "Mom, when asked what you missed, why didn't you say a red Corvette!"

We unpacked a little more. I found some new books. Talk about feeling rich! While continuing to unpack, I was mentally occupied by trying to decide which to read first. Yuri Usachov dug through the Progress until he found the boom box he had ordered. Once again, we stopped work, this time to discuss where we should mount this tremendous gift. Once we decided on the Base Block, Yuri engineered a shelf, to which we strapped the boom box before loading it with CDs. From then on, we had almost continuous music in the Base Block. We were also able to listen to it in Priroda by using Mir's intercom system. Almost immediately, it was hard for us to imagine what our life had been like before we had this capability of background music.

Yuri Onufriyenko chose our first music, Ukrainian ballads. This was a foreshadowing of the background music during my stay on Mir. Most of the music we listened to tended to be Russian or Ukrainian folk music. One evening at supper, Russian folk music was playing in the background, and Yuri Onufriyenko, noticing that I had not yet selected

Tumbleweed

any music, suggested that I put in one of my CDs. I hurriedly floated into Spectra and grabbed the first CD I felt in my bag. It just happened to be a CD of Schubert symphonies selected by my daughters. I inserted it into the boom box, and symphonic music filled the Base Block—for at least two minutes. Then Yuri, with a very pained look on his face, reached over and switched it back to the Russian folk music. His only comment was, "Please, no more Schubert."

Yuri Usachov and I could not stop laughing. As soon as one of us started to get our laughter under control, the other would sputter, "Please, no more Schubert." That would start us both off again. For the rest of the flight, whenever things got a little quiet around the supper table, Yuri Usachov would turn to me, or I would turn to him, and say, "How about a little Schubert with supper?" Yuri Onufriyenko's accompanying bewildered look would always elicit the appropriate Ukrainian joke (Ukrainian jokes are related to Polish jokes) from one of us, with even more laughter.

We loved the fresh tomatoes, fresh apples, and fresh oranges the Progress had brought. But I must admit that I never knew why we were sent so many lemons. Occasionally, Yuri Onufriyenko would cut a lemon and ask us if we wanted a slice. We always turned him down. After all, we had no sugar on the station, so we could not even sprinkle them with sugar if we had desired to suck them for fresh "lemonade."

I rapidly discovered the great error I had made by requesting that my personal psychological support package consist mostly of books and other printed matter. Not only could the contents not be eaten but I could not share them with Yuri and Yuri after a meal or when we had tea. Both Yuri and Yuri had received packets of special candy, cookies, cheese, and little cakes. These were put away to be shared on various occasions with the entire crew. I must admit that I turned down my portion of the big slab of raw bacon, a special Ukrainian treat, Yuri Onufriyenko had received. I would not have been able to eat it under the best of circumstances and especially not after picturing it sitting loaded in the Progress out there under the hot Baikonur sun. Even though I did not eat the bacon, I often wished I had something to share with my generous crewmates.

Progress II

Because I spent more time on Mir than originally planned, I was present for the arrival of the second Progress. Before it was launched,

Lucid

the Russians informed the Americans that they would make a special effort to get something on it for me. Gaylen asked what I needed to keep me happy through my extended stay. My first response was, "Twinkies, mayonnaise, and the sequel to the book that had been on the first Progress—and in that order, please."

Twinkies were top on my list to Gaylen because they last forever and would be a very nice treat for me to share with Yuri and Yuri. As for the mayonnaise, I thought that if I could get my daughters to buy a plastic bottle of mayonnaise and send it up, it would be a great surprise for Yuri and Yuri and save them a lot of time opening those small packets from the American meals.

I had been overjoyed to see the new books on the first Progress and picked out *The Mirror of Her Dreams* to read first. I selected it primarily because it was not written by an English author in the eighteenth century—not that I have anything against eighteenth-century England, but by that time, I just needed to get out of old England for a while. I also knew it was a favorite of my daughter's. I read through it with mounting suspense, reaching the last page just as the hero jumped through a mirror with the villain in hot pursuit. That was it. There were no more words. I floated there, alone in Spectra, in stunned disbelief, holding only volume one. I was stranded, the impossibility of running to the local bookstore forefront in my mind.

For the first time on Mir, the reality of being totally cut off from Earth hit me in the pit of my stomach. How could my daughter have done this to me? Who would send only volume one of a two-volume set to her mother in space? Several email exchanges later, she promised, on pain of total banishment from the family for life, to try to get the second volume on the Progress about to launch.

We eagerly awaited the second Progress of our stay. We knew it would primarily contain science equipment for the CNES mission (CNES is the French space agency), but we also knew it would have at least some of the much-needed supplies for Mir. We were getting low on all supplies except food. A few weeks back, Yuri Usachov had flown up to me after his exercise period and asked if I knew where to find more of the "soft" towels. I told him that there were no more and, for the first and only time, saw his face fall in uncharacteristic disappointment. We were down to only fourteen of the other type of dry towels. I also counted all the small squares of wet gauze we used for toilet paper and informed everyone that from now on, we could

Tumbleweed

only use one each day. There would be none left over for cleaning our spoons and can opener! I also had counted the sets of Russian T-shirts and shorts left and told everyone we could only change every other week until we got new supplies on the Progress.

While awaiting the arrival of our second Progress, I had planned out my reading schedule. I would not quite be finished with *Middlemarch* before the Progress arrived. We had also just polished off the last chocolate candy bar that Yuri had dug out of the bottom of the refrigerator. True, it was a little moldy, but thinking "penicillin," we just scraped it off and enjoyed our small share during our evening tea break.

Launch day arrived, but there was no launch. It had aborted at forty-five seconds. The Progress could not possibly launch for another week. All week, while we waited for the second launch attempt, I kept thinking of chocolate candy melting inside the Progress as it sat atop the rocket under the scorching desert sun. I also imagined the rotting tomatoes, apples, and oranges. Yuri and Yuri told me the story of the time someone had tried to launch a watermelon as a surprise to a crew. Watermelons do not take kindly to launch loads. The crew experienced that firsthand when they opened the hatch and viewed and smelled the exploded melon! I tried to keep my thoughts focused on the books that were in the Progress. The sun could beat down on them all it wanted, and they would remain unharmed!

And then the news came that the Progress had launched. I started to speed up my reading of *Middlemarch* so I would be ready to start my new book as soon as it arrived. (Sometimes I have thought of life as a series of lurches from one great anticipation to the next, jutting out of the Plain of Life. And of all the great anticipations that I have lurched toward, opening the hatch of the Progress ranked right up there with the greatest. It is funny how the perceived anticipation heights are all relative to the daily living field!)

Because this was our second Progress delivery, we expertly searched for our personal bags as soon as we had the hatch open. I rushed to unzip mine, only to find no book, no mayonnaise, and no Twinkies. Suddenly, it was hard to keep the big picture in mind. I was so disappointed, so utterly frustrated, that I just wanted to cry. It was already late, and the next day was fast approaching, so I quickly said good night, turned out the lights in Spectra, and told myself, "Shannon,

Lucid

this is not the end of the world. You should be home in six weeks!" I rearranged my reading schedule in my head as I drifted off to sleep.

The next morning, we got busy unloading the Progress. I was desperately trying not to think of missing Twinkies or books. This Progress seemed even more jumbled than the last one. It was easy to identify the CNES equipment because it was all packed neatly in sky-blue kits. We pulled out package after package of wet towels. We were already swimming in wet towels; we had asked for dry. We already had to take the wet towels and hang them out for a couple of days to get clean, dry ones. This didn't work well because the high humidity on Mir made the towels take forever to dry while adding to the humidity problem. In addition, the towels were impregnated with a chemical that helped clean your body of accumulated sweat and grime, but it left a residue when we dried the towels that made them stiff and nonabsorbent. On the next comm pass, Yuri Usachov asked the ground why they sent up all those wet towels when we had specifically said it was dry that we needed. We received no satisfactory answer from the ground.

As we dug through the Progress, I kept coming across several copies of the *Atlantic Monthly*. They had just been stuck haphazardly into various nooks and crannies. I laughed when I saw them. I could just see what had happened back home. My daughters must have just flown into the house only to hear Mike announce, "I have just received a call that things for your mother need to be delivered to NASA in a couple of hours. Someone is flying to Russia tomorrow and will hand-deliver them to the folks so that they can get them on the next Progress." My two daughters would have looked at each other, shrugged, muttered something about there not being time to get to Half Price Books, and grabbed a stack of my magazines that had been accumulating in a corner over the months that I had been gone.

Yuri Usachov found a present in his bag for each of us: three cloth patches, each individually embroidered with a different and unique design. Yuri explained that they had been crafted with love by his friend who worked in mission control. Mine was a beautiful twelve-inch circle with Earth's cloud-covered ocean represented in the bottom two-thirds and the black star-studded sky dominated by a shuttle in the top one-third. This circle was framed with Russian and American flags. Three large tulips were superimposed on the oceans, and my name was embroidered on the top. I was really touched to

Tumbleweed

think that some unknown Russian woman had spent many long hours designing and then embroidering this for me as a gesture of friendship.

I found a poem from my daughter, and yes, this poem made up for the aggravation engendered by the missing-book fiasco. I made a paper frame for the poem and taped it to a panel in Spectra.

When I Gaze Up

Tonight, when you gaze down, you'll see
a thousand different lands.
And on a hundred different shores
you'll see oceans reaching sand.
But one small light will catch your eye,
not allowing it to roam.
For even from the spires of space,
You can see the hearth of home.

Tonight, when I gaze up, I'll see
a thousand different stars,
Each one a dream, a world unknown,
a promise from afar.
But only one will know my name;
that one I strain to see.
For only one, when I gaze up,
is gazing down at me.

It was Sunday, so each of us was scheduled for a family telephone call. I noticed that mine was set up to use the Wallops, one of the US ground stations. This certainly did not give me any reason to anticipate the call because Wallops never worked. I wondered, to myself, why the ground even bothered to schedule it. At the appointed time, I tried for ten minutes to hear something over the radio but heard only an occasional phrase from a police scanner. While I was doing this, Yuri and Yuri were in the same area going through their bags. In disgust, I removed my headset. Before tiny tendrils of self-pity could even sprout, Yuri Onufriyenko exclaimed, "What is this? Why do I have this? I can't read English."

I looked, and there it was! My book! Talk about pure bliss! Then he pulled a few more things from his bag that were obviously mine. He held up a photograph that he was puzzling over. It was a picture of my friend Ellen's two daughters with a great note on the back, written by Ellen. Evidently, Yuri's bag was not quite full, so the overflow from mine had been stuffed into his. Suddenly, life sparkled. We broke out a new package of cookies and started munching on them with our tea. Then it was time for another comm pass. Down in the SUP (the Russian mission control center), Bill had worked a miracle. He had talked the Russian control center into giving the pass to my family because the Wallops pass had not worked. Life didn't just sparkle; it was so dazzling, I needed sunglasses!

I found out later that the Twinkies and mayonnaise had made it to Russia (I shouldn't have doubted my daughters), but the Russians would not allow them to go on the Progress because each one was not individually marked with an expiration date. "What do you mean, no expiration date!" I exclaimed in disbelief to Gaylen. "Twinkies have so many preservatives in them, they can never spoil!"

I told Yuri and Yuri what I had tried to do for them. As we were munching pickles out of the glass bottle that they had gotten in the Progress, they asked me whom NASA spoke to about getting the Twinkies loaded onto the Progress. Laughing at my response, they told me NASA's mistake had been to go through official channels. Russian crews knew the Baikonur contacts to get the truly important things on board unofficially.

That night we stayed up late, reading and sharing notes from home, talking, eating, and drinking tea. There are times when life is just *so* good! And it takes so little. A call from home. A long-anticipated unread book. A small Hershey's Kiss.

PIRODA

Mir was of a modular design and was assembled in space, piece by piece, just like so many Lego blocks. The first part, the Base Block, was launched in February 1986. Kvant 1 launched in 1987 and docked to one end of the Base Block. Kvant 1 had a docking port for either a Soyuz or a Progress vehicle. The transfer node was at the other end of the Base Block. This served the same function as a hallway in a house, but instead of being a long corridor with doors, it was a ball with six hatches. Kvant 2, launched in 1989; Kristall, launched in 1990; and Spectra, launched in 1995, were each docked to one of these hatches of the transfer node. For a crew person to go from the Base Block to any other module, with the exception of Kvant 1, they had to travel through the transfer node.

The arrival of the last and final module, Priroda, would mark the end of the construction of Mir. Priroda, which is a Russian word that means "nature," was originally planned to dock to Mir during the month before I arrived, when just Yuri and Yuri were on the station. The original plan had it reconfigured and up and running by the time I arrived on Mir, but it was not completed on time, and the launch date slipped. To me, this was a bit disconcerting because not only was Priroda the laboratory in which I was to work while on Mir, but it also contained most of the hardware I was to work with. The bottom line was no Priroda, no work. To say that, after my arrival on Mir, I looked forward with keen anticipation to the launch of Priroda was a gross

Lucid

understatement. Even at night, visions of days jammed full of productive activity danced in my head.

Docking

Finally, the day came when mission control informed us that Priroda had successfully launched from Baikonur and was headed our way. Yuri and Yuri quickly put together the TORU system, a control panel on a pole that was stuck into the floor of the Base Block, right near the command post. If something went wrong with the automatic docking system, Yuri Onufriyenko would use the TORU to manually dock Priroda with the hand controllers and the data presented on the screen. I laughed to myself when I saw the assembled TORU because it looked for all the world like a new Skiffer crew member, a robot standing there ready to serve us. After Yuri and Yuri checked out the TORU system, I figured we were ready for the big event—all that was left for us to do was to just continue floating in a standby mode.

Later that day, the control center surprised us with the information that while monitoring the flight of Priroda, they had received indications that there had been a fire, caused, they thought, by an electrical short. This left Priroda with only half of its electrical power. Concerned, the ground told us that they did not know exactly what to expect during docking. The next thing they told us was for Yuri Usachov and me to be inside the Soyuz at the time of docking. To me, this sounded like the ground thought there might be a hard impact instead of a normal docking!

Soon, it was docking morning. This was the first time something new had happened in six weeks, and we were ready. During breakfast, we started a contest to see who could spot the arriving vehicle first. We discussed what our orientation would be and which window would provide the best view of Priroda's arrival. I rapidly had enough of all such theoretical discussions and took the more practical approach. I just flitted from window to window.

Finally, there against the black-velvet sky, framed by the scattered multitude of random pinpricks of light, a new, faint light appeared. Could it be Priroda? I held my breath and watched. Was it enlarging, was it moving, or was it just my imagination? I left the window and flew to the Base Block, grabbed my package of dried fruit left over from breakfast, ate it, and then went back. Yes, the new light was still there. It was getting bigger. It had to be Priroda. Excitedly, I pointed it out to Yuri and Yuri. All morning, we watched as it came closer and closer, agonizingly slowly. Then it was no longer a star but a space vehicle, a gray bullet speeding directly to the heart of our home. Yuri Onufriyenko was stationed at the TORU, Yuri Usachov was busy filming the approach with the movie camera, and I was trying to be helpful by just staying out of everyone's way.

Yuri Usachov and I were like two unsupervised kids turned loose at the circus. We frantically flew from window to window—Yuri with the big beta camera slung over his shoulder—jabbering with excitement at the latest developments just outside. Yuri Onufriyenko, like the good commander he was, maintained the big picture in his mind. He hollered to both of us to get between him and the Soyuz, which was docked at the end of Kvant 1. Yuri Usachov, a space veteran, explained that he was going to take pictures of the docking out of the window in Kvant 2. He assured Yuri Onufriyenko that he would be able to be in the Soyuz within seconds of a collision. I, also a space veteran, but only of short flights, not long flights, smiled and said I had no problem with sticking close to Yuri Onufriyenko between the TORU and Soyuz.

Priroda slowly waltzed closer and closer. I alternated between looking out the window beside Yuri Usachov's sleeping bag tied to the wall and at the TORU display, which showed a fuzzy image of the front end of Priroda.

Suddenly, Yuri Onufriyenko looked straight at me, stuck his arm out with his finger pointing down the aisle of the Base Block, and shouted, "*Korabel*," the Russian word for "spacecraft."

I smiled sweetly back and nodded an affirmation. Yes, I knew there was a craft out there.

Exasperated, Yuri did a repeat performance.

This time I acknowledged with a Russian sentence, "Yes, I see the spacecraft."

Yuri shouted a third time. Little light bulbs flicked on in my head, and I finally realized that he was not informing me about the rapid approach of Priroda but, rather, ordering me to the Soyuz! Immediately, I executed a flip in midair and shot into Kvant 1 and then into the Soyuz. Better having a happy commander than watching a docking was my philosophy!

I floated in the dimly lit Soyuz and maneuvered my body to see out of its small porthole. No matter how I twisted or turned, I could see none of the action. Then I thought I felt a small bump. Docking? I waited. No one else joined me in the Soyuz, so I figured that the docking must have been normal. After another minute or two, I peered out around the hatch. There were no bodies flying toward the Soyuz, so I slowly floated out and down the length of Kvant 1 and peered into the Base Block. I could see Yuri and Yuri at the other end, working on the hatch behind where Priroda should be. I slowly floated up behind them and listened. They turned and told me that it had been a perfectly standard docking, but we had to wait to open the hatch. The ground wanted us to check the atmosphere carefully before opening it to ensure that the fire had produced no harmful gases.

I was rapidly becoming an expert on how to hurry up and wait! I had been suffering from the illusion that as soon as Priroda docked, we would fling open the hatch and there would be all the new toys for me to interact with. Realistically, I knew that was not going to happen, but my emotional desire would not listen to reason. Behind that dull-green hatch was the biggest Christmas present yet, my own private laboratory—I liked to think of it in this way because, after all, it was filled with the US science experiments—and no one was tearing at the wrappings!

Some Assembly Required

Before we could do anything, we needed instructions from the ground on determining the level of hydrogen sulfide, a potentially harmful by-product of the fire that had occurred in Priroda. The ground eventually told us where to locate the glass tubes that we could connect to a port and then watch the chemical inside these tubes to see if it changed color. In this way, we could determine if the atmosphere inside Priroda was contaminated. Yuri Usachov carefully carried out the procedure while Yuri and I crowded around to watch the tube. I held my breath, hoping we would see no evidence of a chemical reaction. If the atmosphere inside Priroda was contaminated, who knew if we would ever open the hatch!

We watched for a few minutes without the slightest evidence of a color change. That was good enough for us. Time to open the hatch! Yuri and Yuri started to work on the hatch, and I floated just behind them, capturing the big moment on film. Finally, it swung open. I had a momentary sinking feeling. There was no bag inside the hatch containing either fresh fruit, books, or mail from home. I knew there was not supposed to be any; those bags were for the Progresses. But if I had been the person on the ground in charge of closing the hatch on Priroda, I would have slipped in something of a personal nature for the crew!

I looked at the interior of "my lab." I could not figure out how we could even enter it. It was literally wall-to-wall stuff. Every panel had a huge metal frame bolted to it. Each of these frames had big pieces of equipment bolted to them. The "floor" could not be seen because over a hundred batteries, the type used in Russian submarines, were bolted to it. These batteries, instead of solar panels like those of Progresses, were used to power Priroda on its flight to Mir. I hoped Yuri or Yuri knew where to begin, because I had no clue.

Before we could think about how to reconfigure Priroda, the small consideration of power had to be taken care of immediately. Because of the fire after launch, the power to Priroda was in a critical state, so critical that if immediate action was not taken to correct the problem, Priroda would be lost. The ground began to frantically bark out corrective actions to Yuri and Yuri. The Yuris, just as frantically, carried out these instructions. At the next comm pass ninety minutes later, we listened to the hurried, staccato set of instructions. This went on all day. There was no time for Yuri or Yuri to explain to me what was going on. Listening to the comm was enough to convince me that we had a serious problem. It was easy to figure out that Yuri and Yuri had no time for questions, so I just stayed out of the way. Finally, at the end of the day, I could tell by their faces that they had saved Priroda.

I had hoped that completing this first step would enable us to begin getting Priroda up and running, but the ground now had concerns about the possibility of contamination from the fire. They wanted each of us to cover ourselves in long white underwear and don surgical masks and eye goggles before starting to work in Priroda. This outerwear would help protect us from any caustic contaminants in the atmosphere or residue on surfaces that had been caused by the fire or the batteries. These contaminants could burn our skin if we came into contact with them or burn our lungs if we breathed them in. If they permeated the mucous membranes of our noses or eyes, the tiny bits of battery contaminants could blind us.

Lucid

Becoming a Crew

Once in their "surgical gear," Yuri and Yuri started to work, dismantling the metal framework, bagging the batteries, and in general getting Priroda looking like a laboratory. I asked what I could do to help. "Nothing," they said. "We will get it all ready for you, and then you can start to carry out the American science program."

I laughingly told them that was not a good idea. I explained that I liked to work and didn't just want to watch. Again, they told me that it was their job to get it ready for me. I explained that we were a crew and that a crew works together, doing both the good and the bad. A real crew does not exist if one person does only the good stuff and does not help with the dirty work also.

My arguments left them unconvinced. They started to work. I had been in similar situations before in my life and knew that if the "big boys" would not let girls climb trees with them, all you had to do was hang around and be useful, and before you knew it, you would be hollering and climbing and swinging through the branches with the best of them. That is just what I did. I hung out in the hatch, and when Yuri needed the big screwdriver, I had it ready to hand to him. When he had a battery unbolted, I had the plastic bag held open and ready. Soon, we were a seamless relay team of three, unbolting batteries, double-bagging them, and tying them up.

As we worked, Yuri Usachov and I started in with our Ukrainian jokes. Yuri Onufriyenko was very proud of his Ukrainian heritage. Because he was younger than both of us and he was the commander, it was only natural that Yuri Usachov and I teased him at the appropriate times. For instance, when I could not budge a bolt, one of us might say, "Oh, but this is a tough bolt. It is in way too tight for a weak American woman to budge. It is a Ukrainian male bolt." Laughing, Yuri Onufriyenko would obligingly, with one mighty twist, get it moving so that I could finish up the task. The Ukrainian male jokes were a continuing theme for the rest of our flight together.

After we had all the batteries bagged, we began removing all the supporting framework attached to the walls. An observer would have found us a strange sight. Our long white underwear was pulled on over our clothes. Some of us wore our masks, but because they were hot and it was difficult to talk while wearing them, the masks often just dangled around our necks. Some of us wore safety glasses, whereas

Tumbleweed

others didn't, but we all carried big screwdrivers or wrenches as we floated amid the long metal beams. I floated astride a bundle of unbolted metal pipes and tubes, which I had wrapped my legs around while trying to tie it up with cord. The tubes banged into each other, chiming in with different tones, producing the sound of a city at noon, with every cathedral pealing out midday praises. To us, in our sensory-deprived state, it was truly beautiful. As more and more of the metal was bundled up and tied down under the big net that stretched across the floor of Kvant 1, I wondered just how we were going to get rid of all this excess metal. It seemed to me that we certainly had enough metal to build a brand-new space station.

In two days, we had all the batteries removed and bagged, all the metal structure removed and bundled up, and all the new equipment stuffed behind panels in the various modules. We were ready to begin work in Priroda. When Yuri Onufriyenko informed the ground of this, they congratulated us on the great job we had done and told us that they had planned on us taking six days to do what we had done in two. Yuri Onufriyenko then said, "You need to also thank Shannon. She was a great help, and we would not be finished without all the work that she did." As he was saying this, he smiled at me, and I smiled back. We were indeed a crew, working together, no matter what was presented to us, in all phases of flight.

As a reward for completing the reconfiguration so quickly, we were given the next day off as a holiday. My immediate thought was, "Oh no! Not a holiday! We worked so hard to get this done so that we could start to work!" I had already had six weeks of semiholidays and was ready for an all-out work schedule!

My Personal Space Lab

Bill told me which pieces of NASA equipment I needed to check out first. In Priroda, I looked around, trying to figure out where to start. (And no, I had no written procedures to follow.) I opened the panels and saw that none of the wiring exiting from the NASA equipment was connected to any source of power. I did not even know how to describe what the problem was to the ground to get help. So I thought about the apartment I had lived in when in Star City for the last year and how I'd had to familiarize myself with unfamiliar aspects of the Russian electrical system after I had blown out fuses using my American hair dryer. Looking around Priroda, I wondered, "If I were

Lucid

Russian, how would I have put this together?" I told myself to stop thinking like an American and to not expect things to be laid out like they would have been on the shuttle.

Once I was in the right frame of mind, everything began to fall into place. Behind a panel, I found a bundle of wires. By strategically snipping at the restraints on the bundles, I was able to get wires that I could string behind the panels and establish connections to power up all the NASA equipment. When finished, I flipped the switch, and when the appropriate light-emitting diodes (LEDs) lit up, I felt a real sense of accomplishment. I had figured it all out. We were up and running and ready for work.

While I was busy stringing electrical wires behind the panels of Priroda, Yuri Onufriyenko floated in. After watching for a few minutes, he asked, "Shannon, did you receive much training on Priroda in Star City?"

Surprised by the question, I looked at him and said, "No, I never received any training on Priroda. Didn't you and Yuri get a lot of Priroda training?"

He laughed and said, "The only training we received was the same you did. Remember that night when we went over to the factory and, for forty-five minutes, walked around the module there on the factory floor? That was the only training we received."

I was a little taken aback by this revelation. I had been unwittingly thinking like an American again and had assumed that the masterful way in which Yuri and Yuri were reconfiguring Priroda was due to extensive hours of training on mock-ups that I had never seen.

The next day when I went into Priroda, I heard a barely audible tone. I could not figure out what it was or even if I was actually hearing something real. Later in the day, I asked Yuri Onufriyenko to come in and listen. He agreed with me that there was a faint noise. And then comprehension spread across his face, and he said, "Oh, that is the fire alarm! It must have been set off by the smoke detector."

"You mean to tell me that I have been listening to the fire alarm all day and didn't even know it? I don't see or smell any smoke. What has set it off?"

We opened a panel near the smoke detector and pointed my flashlight into the dark crevice between the panel and the hull. We were astounded by the infinite number of tiny particles we saw dancing in the beam of light. We deduced that these particles were created when

Tumbleweed

I had cut apart the wrapped wire bundles. We never were able to raise the volume of the fire alarm loud enough to catch the crew's attention if there had been an actual fire.

On Mir, I was living every scientist's dream. I had my own brand-new laboratory in which I could work independently for most of the day. Before one experiment had time to become routine, it was time to start on another with new equipment and in a new scientific field. I discussed my work at least once a day with Bill or Gaylen, both at the Russian mission control. They coordinated my activities with the principal investigators, the American and Canadian scientists who had proposed and designed the experiments. Often when we started a new experiment, Bill arranged for the principal investigators to be listening to our radio conversations so they would be ready to answer any questions I might have.

My role in each was to do the appropriate procedures. Having procedures that I could use sounds straightforward, but it certainly was not. Before every new experiment was started, Bill arranged to have access to the training hardware for that specific experiment. Using this hardware and the Russian procedures I had on board, he then painstakingly went through them to thoroughly understand what the experiment was to accomplish and then translated them into English and wrote them in the NASA procedure format. Then he sent the English procedures to me via the ham radio packet system. If Bill had not been so focused on ensuring I had good procedures on board, the NASA science program for my flight would not have been the resounding success that it was.

The data and samples I collected were returned to Earth with me on the space shuttle and sent to the principal investigators for analysis and publication. My experience on Mir clearly showed the value of performing research on manned space platforms. During some of the experiments, I was able to observe subtle phenomena that a video or still cameras would have missed. Because I was familiar with the science in each experiment, I was able, in real time, to modify the procedures as needed. Also, if there was a malfunction in the scientific equipment, I was present to fix it. Only one of the twenty-eight experiments scheduled for my mission failed to yield results because of a breakdown in the equipment.

Lucid

Quail Eggs

I started my work on Mir with a biology experiment examining the development of embryos in fertilized Japanese quail eggs. The fertilized eggs had been brought to Mir on the same shuttle flight that had brought me, then transferred to the Russian incubator on Mir. Over the next sixteen days, I removed the thirty eggs one by one from the incubator and placed them in a 4 percent paraformaldehyde solution to fix the developing embryos for later analysis. Then I stored the samples at ambient temperature.

Although this sounds like a simple experiment, creative engineering was necessary to accomplish the procedure in a microgravity environment. NASA and Russian safety rules called for three layers of containment for the fixative solution; if a drop escaped, it could float into a crew member's eye and cause severe burns and possible blindness. Engineers at the NASA Ames Research Center designed a system of interlocking clear bags for inserting the eggs into the fixative and cracking them open. In addition, the entire experiment was enclosed in a larger bag with gloves attached to its surface, which allowed me to reach inside the bag without opening it.

Investigators at Ames and several universities analyzed the quail embryos at the end of my mission to see if they differed from embryos that had developed in an incubator on the ground. Remarkably, the abnormality rate among the Mir embryos was more than four times higher than the rate for Earth-incubated control embryos. The investigators believe two factors may have increased the abnormality rate: the slightly higher temperature in the Mir incubator compared with the ground-based one and the much higher radiation levels on the space station. Yes, radiation exposure was high on Mir. Other experiments determined that my average radiation exposure on Mir was the equivalent of getting eight chest X-rays a day.

Growing Wheat

I was also involved in a long-running experiment to grow wheat in a greenhouse in the Kristall module. American and Russian scientists wanted to learn how wheat seeds would grow and mature in a microgravity environment. The experiment had an important potential application: growing plants could provide oxygen and food for long-term spaceflight. Scientists focused on the dwarf variety of wheat

Tumbleweed

because of its short growing season. I planted the seeds in a bed of zeolite, an absorbent granular material like kitty litter. A computer program controlled the amount of light and moisture the plants received. Every day, we photographed the wheat stalks and monitored their growth.

At selected times, we harvested a few plants and preserved them in a fixative solution for later analysis on the ground. One evening, after the plants had been growing for about forty days, I noticed seed heads on the tips of the stalks. I shouted excitedly to my crewmates, who floated by to marvel along with me at seeing the tiny seeds. John Blaha, the next American astronaut to live on Mir after me, harvested the mature plants a few months later and brought more than three hundred seed heads back to Earth. But scientists at Utah State University discovered that all the seed heads were empty. The investigators speculate that low levels of ethylene in the space station's atmosphere may have interfered with the pollination of the wheat. In subsequent research on Mir, Michael Foale planted a variety of rapeseed that successfully produced viable seeds.

Other Experiments

The microgravity environment on Mir also provided an excellent platform for experiments in fluid physics and material science. Scientists sought to further improve the environment by minimizing vibrations. Mir vibrates slightly as it orbits Earth, and although the shaking is imperceptible to humans, it can have an effect on sensitive experiments. The movements of the crew can also cause vibrations. To protect experiments from these disturbances, we placed them on the Microgravity Isolation Mount, a device built by the Canadian Space Agency. The top half of the isolation mount floats free, held in place solely by electromagnetic fields.

After running an extensive check of the mount, I used it to isolate a metallurgical experiment. I placed metal samples in a specially designed furnace, which heated them to a molten state. Different liquid metals were allowed to diffuse in small tubes, then slowly cooled. The principal investigators wanted to determine how molten metals would diffuse without the influence of convection. (In a microgravity environment, warmer liquids and gases do not rise, and colder ones do not sink.) After analyzing the results, they learned that the diffusion rate is much slower in space than on Earth. During the procedure, one of the brackets in the furnace was bent out of alignment, threatening the completion of the experiment. When I pointed this out to Yuri Usachov, he simply removed the bracket, put it on the workbench located in the Base Block, and pounded it straight with a hammer. This kind of repair would have been impossible if the experiment had taken place on an unmanned spacecraft.

Many of the experiments provided useful data for the engineers designing the International Space Station (ISS). The results from our investigations in fluid physics are helping the space station's planners build better ventilation and life-support systems. The research on how flames propagate in space may lead to improved procedures for fighting fires in space.

Many times I have been asked the question, "What was your favorite experiment you conducted while on Mir?" I always want to be careful in how I answer this one because the principal investigators had put long hours into developing their experiments, and I certainly did not want to slight anyone's work. Some of the experiments were, from the crew person's viewpoint, much more interesting than others. The best ones to work on were those that required a lot of crew involvement—such as making observations and real-time decisions about what to do next. The least interesting experiments were those that were self-contained—the ones that were enclosed in a box that required nothing from the crew other than maybe pushing a button to get them started. This observation should be of no surprise to anyone! These are the same reasons why people who work in laboratories love (or don't love) their work.

The candle experiment provides a classic example of the importance of real-time crew involvement. The basic question of this experiment was simple: Do candles burn the same way in microgravity as they do in a $1\ g$ environment? To answer this question, candles of various thicknesses were inserted into a holder, which, in turn, was placed inside the NASA glovebox in Priroda. (A glovebox is basically an enclosed box in which an experiment can be carried out. It is enclosed so that any contaminants involved in or produced by the experiment can be removed and not enter the outside environment. The investigator's hands, inside a pair of gloves, manipulate the materials.) An electric spark ignited the candles. The Priroda glovebox had ports (small windows) for a video camera and a 35 mm camera as the means of data collection.

Tumbleweed

I set up the experiment. Yuri and Yuri requested that I let them know when I was ready to light the first candle. They wanted to be there to watch. We had little in the way of new events to change our daily routine, so lighting a candle was a unique break, a novel occurrence that we wanted to experience together. After the second comm pass of the morning, we all gathered in Priroda and positioned ourselves around the glovebox. We lit the candle and watched it burn.

The flame was different from a candle flame on an Earth birthday cake. Instead of a pulsating orange flame, ours was a blue cap of light sitting just over the wick. It looked like a translucent blue igloo shimmering in the Arctic night. The candle burned on and on. It burned much longer than had been predicted by the principal investigators. Once the candle had been consumed, the blue igloo disappeared.

I rewound the video and played it back. Nothing could be seen. The blue of the flame was not detected by the video camera. Quickly, I found the colored pencils I had on Mir and rapidly drew pictures in my notebook of the burning candles. At the conclusion of one data-collection run, I just happened to turn on the lights inside the glovebox before a candle had completely gone out. Immediately, when it was extinguished, the chamber filled with a puff of gray that looked exactly like a dandelion flower gone to seed. This was totally unexpected. This was something that I could catch on the video! So that is what we did. We improvised a way to video the candle runs as the candles extinguished. This was an exciting time for the three of us because we were seeing what was happening and then maximizing the way we were collecting information to obtain data that the investigators needed to answer fundamental questions about how flames propagate in space. And best of all, our observations led to even more questions!

This was just one small example of the many ways we had to improvise on our own for the experiments to be successful. The reason why this flexibility turned out to be so crucial was that we did not have good comm with the ground. So, because the time was very long between comm passes, we often made our own changes to the way we were conducting the experiment.

Many times, while working in my Priroda laboratory, I thought back to the instant when I first decided that I wanted to be a scientist when I grew up. In the fourth grade I found out that water was made up of two gases, oxygen and hydrogen. This fact, that two gases could get

Lucid

together and form a liquid, water, just filled me with wonder and awe. It amazed me that a human, a chemist, had figured this out. I did not know how the chemist had figured it out, but I knew then and there that I wanted to grow up and make such exciting discoveries! Working in Priroda was a fulfillment of the childhood dream of "making discoveries," but who would have ever dreamed that the setting for this action would be a laboratory orbiting Earth every ninety minutes!

And maybe, just maybe, working there in Priroda, an overarching life principle was revealed to me. The principle that maybe, just maybe, we don't realize our dreams have indeed come true because of the "undreamed of" manner in which life plays out!

EXTRAVEHICULAR ACTIVITY (I.E., SPACEWALKING)

The evening came when Yuri Usachov casually made the statement, "Now we are going to start a very interesting time here on Mir." As the next five weeks would show, he had, in his usual understated way, made a very truthful statement. We were indeed starting a period when our way of life could be summed up as, "The start of another week, time to prepare for another extravehicular activity (EVA)."

The rhythm of those five weeks stayed the same; only the specific tasks varied. When Yuri made his announcement, I did not have a clue as to what performing EVAs from Mir would entail. In Star City, I had no training for being the intravehicular activity (IVA) crew person, the crew person left inside the space station when the other crew members were outside doing the spacewalks. I quickly got a feel for what was going on while preparing for the first EVA.

For about a week before that first EVA, Yuri and Yuri spent their days checking out their space suits and preparing the airlock. I was amazed at the versatility of the Russian suits. They were designed to be serviced by the crew members aboard Mir because there was no way for them to be serviced by ground personnel. For instance, at the completion of their fifth EVA, Yuri Onufriyenko informed me that the suit he had been using had now been used for thirteen EVAs, all with no ground refurbishment! In addition, the suits were also designed to accommodate crew people of varying sizes.

Lucid

Yuri and Yuri spent a lot of time floating around the table in the Base Block, choreographing the time they would be outside on the surface of Mir. The lack of procedures from the ground surprised me. I'd had a lot of time in the American space suit doing simulations in the Weightless Environment Training Facility (WETF), the gigantic swimming pool at the Johnson Space Center where astronauts trained for spacewalks before the advent of the International Space Station training. Every run was dominated by procedures. Every move was choreographed by the procedure writers and preplanned. This was certainly not the case in the Russian system. I asked Yuri and Yuri how much time they had spent in preflight training for their upcoming EVAs. They answered that they did just the basic EVA training, not specific training for each task like we did at NASA. They asked me how much time I had spent in the WETF at NASA, and when I told them an estimated number of hours, they were flabbergasted. They basically had done a generic training flow so that they would know how to use their suits. They had completed only one flight-specific training exercise, and that was for the installation of the French experiment package, probably because the French had paid for the training time. In addition, a videotape arrived on our first Progress, which was from their training run on this French payload. This tape was for them to review before they went outside to do the installation.

Another surprise came when I realized from the Form 24 that we were to take a nap in the middle of our day and that the EVA was to be done during our night. When I asked about this, I was told that we were doing it at night because, due to the orbit that we were currently in, radio coverage was better at that time. The ground wanted the best radio coverage while Yuri and Yuri were doing their EVA. I understood the part about wanting good radio coverage, but why not just wait a few weeks until our position had changed once again and we were getting the best ground tracks over the Russian ground sites during our day? The answer was simple. Now was when the ground wanted to do the EVAs. I meditated awhile on how different the Russian philosophy was from NASA's. At NASA, everything was done to optimize the situation for the crew. The Russian system appeared to try to optimize the situation for the ground!

As the time approached for the Mir 21 crew to do their first EVA, I floated into Spectra, turned off the lights, and to my surprise, fell asleep for a several-hour nap. After the alarm woke me up, I drifted

Tumbleweed

into the Base Block. Yuri Onufriyenko greeted me by saying, "I need to train you for what you will be doing while we are outside." (I silently wondered why I was getting this training just hours before the actual EVA.) "Ground wants you to periodically input commands to change the orientation of Mir while we are doing the EVA."

"OK, but you need to make sure I know what I am to be doing. I certainly don't want to start Mir moving if you are not hanging on! I don't want you spinning off into space and leaving me inside all by myself!"

We floated over to the main computer panel, and Yuri started to rattle off the commands that I was to execute at various times during the EVA.

"Wait. I need to write this down!"

"No, you don't need to write anything down; I am telling you what you are to do! You just need to listen."

"You don't understand. I want the procedure down in writing so that I don't make any mistakes!"

"You don't need anything in writing. Just remember what I am saying!"

"I am just an old American female! I can't remember things as well as you can. I need a written-down procedure."

"OK, then. Write down what I am saying!"

This conversation was a classic illustration of the yawning gulf between Russian and American space operations. No astronaut would ever execute a procedure without having the written version in hand and checking off each step as it was completed. The Russian cosmonauts routinely executed verbal procedures from the ground, never writing anything down. In thinking about this, I realized that it highlighted the differences in our educational experiences. Educated in America, I relied heavily on the written word, whereas Yuri and Yuri, educated in Russia, relied predominately on the spoken word.

Yuri began to issue commands for me to write down. When he started saying, "Igor," I said, "Wait. What is this Igor?" It was then that I stumbled onto the obvious—that the Russians, just like the Americans, have a phonetic system used in radio communications when saying letters of the alphabet so that there will be no confusion when they are enunciated over a crackling headset. Of course, for years, every time I had gone flying, I had been saying Romeo, Hotel, Papa, and so forth while talking to flight controllers and when using

Lucid

"airplane talk" with my friends. As soon as I understood that the Russians did the same thing, it made perfect sense. The only question I had was why on earth, while on Earth, no one had clued me in to this during training instead of me stumbling upon this realization just hours before I would be issuing commands to the Mir control system for the first time.

I finally had my short procedure for issuing commands to change the orientation of Mir written out. I made sure Yuri watched while I practiced going through them by touching each key on the control panel that I would push during real-time operations. When we were both satisfied that I knew what to do, he said, "If a fire happens while we are out, you need to . . ." Noticing that my eyes were glazing over from information overload, he turned to the wall behind us; removed a curled-up, yellowed portion of typing paper; and said, "Here. What needs to be done in emergency situations is all written out." As I strained to decipher the faded Cyrillic in the dim light, he said, "Oh, time is short. Nothing will happen." He stuck the tattered sheet back on the wall and proceeded with preparations for the EVA. During the next ground pass, he told the ground that I was checked out and ready to carry out whatever commands they wanted done while he and Yuri were outside.

I made a few jokes about both of them needing to be nice to me because I would be in charge while they were out and who knows what surprise might be in store for them when they returned. Maybe I would transform the Base Block into Little America, with the Stars and Stripes flung across all panels. When Yuri, in all seriousness, asked how many American flags I had aboard Mir, I decided I needed to tell them I was joking. Sometimes I just needed to make sure they understood when a joke was a joke!

Then it was time for Yuri and Yuri to don the white underwear to be worn under the Orlan, the Russian EVA suit. (*Orlan* means "eagle" in Russian.) It is basically one-piece underwear, except that it is covered with plastic tubing that has been sewn into place. This tubing is connected to the space suit's water system. Cold recirculating water would remove the heat produced by Yuri and Yuri while inside the space suit. American suits use the same type of system. The major difference is that the Russian underwear has a hood that covers the top of the head to keep the head cool while working hard. In addition, each set of Russian underwear aboard Mir had two small blue bows just

Tumbleweed

below the waist. These bows had been sewn in place by the babushkas who, with loving care, had hand-sewn each pair of underwear. These bows just made the entire outfit look really cute.

Just before it was time to enter the hatch, Yuri Onufriyenko configured the comm system so I would be able to talk to both of them while they were outside and so we could all communicate with the ground. Yuri admonished me not to change the configuration. I laughingly took out the red tape that I always had in my pocket and put a big piece across the switches of the comm system, emphasizing that nothing would be touched. We laughed a little, and that eased some of the tension that was mounting as we awaited the time for both Yuris to enter the airlock and begin their EVA.

The Russians have a long-standing tradition that before any important event, such as leaving for a trip, everyone involved sits for a moment. The purpose is for everyone to have a moment to collect their thoughts (e.g., Did I pack my toothbrush? Do I have my train ticket?), catch their breath, and focus on the task at hand. Yuri Onufriyenko announced that it was time for us to sit together. The three of us floated in the Base Block, not saying anything, hands folded, quietly collecting our thoughts. Too soon, Yuri Onufriyenko broke the spell of quiet peace with the words, "OK, let's go," and the two Yuris literally flew over my head, like two white geese headed south, exiting the Base Block with a wave. They flew through Kvant 2 and into the airlock. I was behind them with the camera, and once inside the airlock, they both turned. As they smiled and waved, I snapped a picture. With a "see you soon, Shannon," they clanged the airlock shut. They then started suiting up. I was alone and the captain of the ship for the next five hours.

Comm was very good between the IVA (me) and the EVA crew (Yuri and Yuri). I heard all the preparations as they were getting into their space suits and then depressurizing the airlock. Every once in a while, they would ask me if I was listening and what I was doing. Sometimes they would ask me what the atmospheric pressure inside Mir was or what part of the world we were currently flying over. When they had a period of waiting, they asked me to place my headphones close to the CD player in the Base Block so they would have music to listen to. Finally, I heard them exiting the airlock and leaving the Mir station.

Lucid

Once they exited, Yuri began yelling at me to look out the window and start taking pictures of the two of them. I looked out, and there was my commander, perched on the end of a very long white pole over the blue-and-white Earth beneath. It looked like a gigantic fishing pole with a big just-reeled-in fish, Yuri, on the end. Because the outside of Mir was such a large structure, this pole, the Strela, was used to transport a crew person and payload from one segment to another. It was swung manually, like a large crane, from module to module by the other EVA crew person. My first thought was that I was witnessing the future of work in space. For a number of years, I had been working sporadically on the planned space station and had seen artist renditions of what the station would be like when assembly was complete, with astronauts working on it in a routine manner. This, however, was no artistic fantasy. This was real life. The "future" was being played out right before my eyes, and I was being allowed to have a small role in it.

Because Mir was huge and the windows were relatively few, I could only see bits and pieces of an EVA. After one EVA when Yuri and Yuri were looking at the video I had taken, they asked why I had only filmed their backs. I told them that I could only capture what I could see! We named that video "Cosmonaut Spines." Although my vision was limited, I could hear everything. As they clunked across the top of Kvant 1, I had to repress the urge to shout, "Get those reindeer off my roof!" as my husband used to do at Christmastime when the kids were young. During the night passes, I watched them work in a small, flat pancake of light out on the end of some module and heard them muttering about the mamas and the papas (the Russians use these terms instead of *male* and *female* for differentiating connectors) as they worked on connecting a payload to station power. It felt warm, cozy, and homey, just the three of us in the universe.

Then it was time for a comm pass. Mission control started rattling off a long string of numbers that they wanted me to input into the computer. I was caught a little by surprise and started scrambling to get my paper out and my pencil in hand to copy them down. Before I could sort out what I had heard, Yuri Onufriyenko, out on the end of the pole, said, "Shannon, listen," and then he started reeling off the entire string to me. I was impressed. I was unable to copy the entire string while the control center was transmitting it, but Yuri was able to repeat it from memory after hearing it only once when he was outside doing an EVA. This is another example of the difference in how Russians and Americans do business. Never in my life have I been responsible for giving back commands without writing them down. (That's why, when flying, air traffic controllers say, "Read it back.") In Russia, that is a way of life. Quickly, Yuri and I established a rhythm of every time the ground told me a procedure to do, Yuri repeated it to me as I wrote it down. Of course, I read it back to him before executing it.

I had my messy hand-scribbled procedure that I had hastily written down just before the EVA. I had my watch. At the appropriate time, I put in the string of numbers and then executed the command. I held my breath and waited. Nothing blew up. I looked out the window. Yuri and Yuri were still attached to the station. So far, so good. I could relax a little until it was time to enter the next set of commands.

The back of my neck was sopping wet, and I wondered how I had gotten into this predicament, executing critical actions, actions with dire consequences if incorrectly done, with no training. I suddenly remembered when I had been in a similar situation many years before. Hadn't I vowed then to never, never let it happen again?

I was a new pilot with less than one hundred hours of flying time. Having no money to fly, I hung around the airport, hoping that someone would say to me, "I need someone to fly my airplane—could you do it?" One day, my fantasy merged with reality. I was hanging around the airport, wishing I had money to fly, when someone I vaguely knew came up to me and asked if I could fly a Cessna Skylane. "Sure!" I replied, undaunted by the fact that I had never actually flown a Cessna Skylane, much less even been on one, before. It was, though, a high-wing airplane, with four seats, and looked a lot like a Cessna Skyhawk, in which I had about twenty-five hours. I knew it was a little larger than the Skyhawk, and it had a constant speed prop, but I ignored these differences in my eagerness to fly.

"Good," the trusting pilot responded. "I need someone to fly with me up to Kansas. I had to leave my plane at the Hutchinson airport last week when the entire state was covered with tornadoes and tornado warnings. We can fly up together, and then you can fly my Skylane home."

"Great! Just one minute!" The other pilot probably assumed I was ducking into the hangar to use the restroom, which I did, but I also

Lucid

took the time to go to the front counter and quickly buy the owner's manual so that I could find out how to operate the Skylane.

As soon as we landed, the Skylane's owner tossed me the keys and said, "See you back at Wiley Post." I scrambled to get out of the plane before he took off. Once in the Skylane, I opened my new owner's manual, studied the takeoff section, and then took off. I flipped the manual to the level-off section at cruising altitude and then set the mixture according to the written procedures. For about an hour, I cruised along, filled with joyous satisfaction. Not only was this the largest plane I'd ever flown, but it was the largest plane I'd ever been in period, except, of course, for my very first airplane ride so many years before.

Eventually, my excitement wore off, and I began to think about the flight's end. How was I going to land this huge machine? The back of my neck became drenched in sweat. I flipped through more pages and planned out the approach and landing into Wiley Post Airport, vowing that if the plane and I made it down in one piece, I would never do such a stupid thing again. I would always train on how to perform critical actions before doing them the first time for real!

For the record, the Skylane and I made it down in one piece.

Before the first EVA, the Yuris joked about what I would be doing while they were outside and I was the "commander" inside Mir— commander by virtue of being the only person inside the station. They jokingly agreed with each other that I would have a large American flag hanging in the Base Block to greet their return. Well, no, I did not hang up the American flag—I was not sure how far to stretch their sense of humor—but I did make one command decision. For several weeks, we had been eating what had been left in the various food containers and not opening any new ones. As you might guess, the selection we had was not any of our favorites—that is why it was left! So being in command, and knowing that the first prerogative of a good commander is the welfare of her troops, and feeling very much like Captain Kirk of the starship *Enterprise*, I decided to open a new food container and have Yuri and Yuri's favorite meat-and-potato dish hot and waiting for them. Eating it with gusto after the EVA, neither one of them asked where it had been found. All they said was, "Thank you so very much!"

Yuri and Yuri were exhausted when they came in from the first EVA, but before we could even discuss the spacewalk among

Tumbleweed

ourselves, ground was on the radio ordering Yuri to turn on, sometime in the next hour, the Electron system, the system that produced our oxygen. After we lost contact with the ground, Yuri asked me if I had turned the Electron off and on during my training in the simulator in Star City. When I said that I had, one time, he repeated the ground's last uplink and told me that I could stay up and do it; he and Yuri were going to bed. Once again, I was in charge of carrying out a vital procedure with no one even watching to see if I was making a deadly error. I breathed a sigh of relief as I finished the procedure with no anomalies, and oxygen started flowing into Mir.

The outline, the rhythm, of the five EVAs remained the same—for instance, we did all of them in the middle of our night—but each one had its own distinct flavor and incidents.

The main purpose of one of the EVAs was to film a Pepsi commercial. Talk about a major culture clash, a major conflict of interest! As a US government employee, it was absolutely verboten for me to endorse a product, and I was absolutely prohibited from taking part in, or appearing to take part in, any such type of commercial activity. And I should know! Every year I had to read the NASA employee ethical guide and sign a slip of paper acknowledging that I had read it and understood what I had read. Ignorance of regulations could never be an excuse. The Russians, though, newly freed from the restraints of communism and freshly exposed to the capitalistic world, were willing to use Mir to endorse just about anything. While I was on board, Hewlett-Packard laptops, Israeli milk, and MTV were all hawked to the spinning Earth beneath. I made sure that I was nowhere near the Base Block when these commercials were being made. The Pepsi commercial, however, was by far the most elaborate.

Yuri and Yuri were well aware of the constraints under which I operated. I sensed their dilemma. They were going to be outside Mir, hanging out on the Strela with their huge blow-up Pepsi can, and they needed to have pictures of this activity sent down to Earth. My job was to video all EVA activity. How could I videotape the Pepsi commercial without breaking NASA regulations? Luckily, a justification occurred to me that might even hold up if I was ever closely questioned by my management as to my activities aboard Mir.

I explained to Yuri and Yuri that I was aboard Mir as a member of the Mir 21 crew. As such, it was my duty to help my crewmates fulfill all their requirements. Furthermore, I certainly needed to carry out all

Lucid

activities that my commander wished to have done while I was on board. My job was to video all EVA activity. I needed to do this for safety reasons. For instance, if something unexpected occurred during the EVA, it needed to be documented. I could not help what appeared in front of the camera when I was performing my prescribed activities. Therefore, I told Yuri and Yuri that I had no problem with filming all the activity that took place during the EVAs.

During the Pepsi commercial EVA, as I was videoing the action, I was very surprised to see a six-foot replica of a Pepsi can pop up on the end of the Strela, along with Yuri Onufriyenko. Yuri Usachov slowly rotated the Strela so that Yuri and the blow-up Pepsi can were in the perfect location for a shot with Mir in the frame and Earth and the black universe beneath. Inside, I was busy setting up the cameras and talking to the ground, flying from one headset to another. As I was focusing the camera, I had to monitor the time carefully so I would know when to make another input to the computer. I was searching for the eyepiece of the small video camera, which had somehow floated off the camera and was now lost inside Mir, when Yuri Onufriyenko yelled into the headset, "We are ready."

"OK, hang on. I'll start the movie camera after I focus," I replied.

Impatiently, Yuri hollered again, "Are you finished yet?"

"No. I'm still working on the focus."

Then I overheard Yuri ask Yuri, "What can Shannon possibly be doing in there? Just how long does it take to focus a camera?"

We started to laugh uncontrollably. I had almost regained my composure, only to lose it again when Yuri Usachov said, with an exaggerated Ukrainian accent, "Just what can Shannon be doing in there?"

I repeated the Russian word that I had quickly made up, *focuseravate*. In the heat of the moment, not being able to remember the actual Russian word for "focus," I simply added a Russian verb ending to the English word. This cracked Yuri Usachov up. He could not stop laughing. Every two seconds, he would ask, "Shannon, are you still *focuseravating?*"

Despite the laughter, I finally got a great focus and a frame that I liked. As the swirly blue-and-white Earth beneath gave way to the African rust Earth, I sent the picture down to mission control. It looked impressive—Yuri Onufriyenko on the end of the pole, the rust-brown African continent spinning beneath him, and the huge blue

Tumbleweed

Pepsi can slowly rotating in front. Yuri looked for all the world like a cowboy of the Wild West trying to lasso the can. Later, I found out that the people in mission control burst into applause when they saw the footage. Sadly, Pepsi never used it. The company's advertising campaign had apparently taken a different direction.

At the completion of each EVA, Yuri and Yuri entered the airlock and began the process of repressurization. After many requests for me to read aloud the station pressure from the big manometer in the Base Block, the airlock would finally open, and they would emerge, looking like two excited young boys returning from a great adventure. They would immediately watch the video I had taken, excitedly discussing each event while drinking the hot tea or the tube of juice that I had waiting for them. After the last EVA, I had what I thought was Yuri's favorite Russian tube of juice by his place, as usual. He grabbed it with a broad smile of thanks, which immediately turned into a horrible grimace as a huge glob of unexpected ketchup squirted into his mouth. Yes, I had mistakenly gotten the wrong tube—well, they do all look alike, and with the paint peeling off, it was hard to read the name. My language skills were not quite at the level where I could have convinced him that I should at least get points for trying!

After we had finished talking about the EVA, it was always off to bed, and after several hours of great sleep, we woke up refreshed and talking about the next EVA later in the week. I often fantasized that maybe this time, the guys would invite me to go out with them—yes, it is always greener on the other side of the hatch!

FORTY-FOUR BAGS OF SOUP TO GO

I had inhabited Mir for one hundred days. I had read one hundred Psalms. I had forty-four bags of soup to go. I was drinking one bag of soup every morning for breakfast and counting them down one by one, day by day, until I would be home again. The finish line for my space marathon was within visual range.

I spent every day working in Priroda on the science program. Work was going at a nice pace. I could definitely see how all the experiments could be completed within the next forty-four days. The ground had a different perspective, however. The science program for this mission had been slow to start because Priroda had been delayed, and the NASA team in the Russian control center was beginning to get stressed out, trying to figure out how everything would be finished in time. They were working in a bureaucratic nightmare, trying to satisfy all the Russian and American requirements.

I was, literally, far above all the paper hassles. Actually, by this time, I was working without any paper at all. At first, all the modifications to the procedures I was doing were coming up via the ham radio packet system. I would print out the message and work from this sheet of paper. Eventually, the paper supply on Mir was finished. Yuri devised a system of putting all the messages received via the ham radio on some floppy disks that had been left by a previous crew. I transferred the procedures to one of the NASA laptops that were on board. I plugged in the laptop near my work area and read the procedure directly from the computer. NASA's dreams of a paperless space station had been realized!

With only forty-four days left to go, I had already set a record. On the 115th day I had spent on Mir, Bill Gerstenmaier concluded his morning transmission with, "Congratulations, Shannon, you now have more time in space than any American." Even though I knew it would be a record for only a short time, it was a record I was glad to have, and I knew exactly why. In Bill's sentence, the word *American* had not been followed by the word *woman*. It was not the longest time for an American woman but the longest time for an American. It was a real record and not just a "pretty good for a woman" type of record.

Upon hearing Bill's comment about my just-set record, I immediately had a flashback to the year 1960. That was when I participated in the National Science Fair in Indianapolis. One of my goals while working on my project had been to not only get to go to the National Science Fair but to win. After the judging was completed, I sat with clammy hands and bated breath, literally on the edge of my chair, at the awards banquet, straining to hear in case my name was called. As I listened to the names being called, I suddenly realized that all the winners were divided into two categories: boys and girls. By the time my name was announced in the group of second-place winners in the girls' division, it meant nothing to me because I was deep in the struggle of trying to understand why my project was considered so unworthy it could not even be judged with the boys' projects.

The evening of my one hundredth day on Mir, I typed an ebullient email to my family, telling them that I was counting down the soup bags until I would be with them again. After sending it, I read the email I had received from home that same day. An email from Shani, who worked for a NASA contractor and hence heard a lot of the NASA gossip, informed me that STS-79 might be delayed because a problem had been found with the solid rocket boosters (SRBs) from the last shuttle launch. (The SRBs from each shuttle flight were reusable, and after they were fished out of the Atlantic, they were closely examined to determine how they performed before being refurbished for the next flight.) This meant STS-79, my ride home, might be delayed, and my four-month expected stay extended to six months. My first thought was, "I should never have started counting those soup bags!"

After reading this note, I immediately started to think about my body. How would it hold up after six months of microgravity? What

Lucid

condition would I be in when I returned home? I had attended a meeting about ten years before, sponsored by the Johnson Space Center life-science folks. The subject was planning for the space station and how long the initial flights could be. The discussion centered around whether the initial space station flights could start at ninety days. The medical community was not happy with this prospect, but they reluctantly agreed, insisting that incremental data on the effect of long-term stays in space be collected and examined before station visits were increased to the goals of four and six months. Now I was staying for six months—all in one fell swoop. No "reams of data" had been collected. No slow, cautious, steady progression of flights had proceeded me. This was a quantum leap! Then I just told myself, "Shannon, don't sweat it. Look at all the Russians who have been in space for six months and much longer. You saw lots of them walking around Star City, happy and healthy. What better empirical data do you need?" I laughed to myself and did not bother thinking about the subject again.

Considering the possibility of my stay being extended, I thought, "No problem—things are good here. I am content. I have good companionship; plenty of food, water, and oxygen; interesting work; and books to read." But then there were those nagging thoughts. "Will things stay this good? Sooner or later, shouldn't Yuri, Yuri, and I start to get on each other's nerves? Wouldn't it be prudent to quit while ahead?"

A month previously, Yuri and Yuri's stay had been extended. Originally, they were to be up for only five months. Because financial considerations were beginning to be the primary driver in the Russian space program, their stay on Mir had been extended to just over six months. This is the maximum time that a Soyuz, the vehicle cosmonauts return home in, can remain on orbit. This extension became the policy of the Russian space program. All the Russian crews after Yuri and Yuri stayed on Mir a little longer than six months. I had seen how Yuri and Yuri handled their extension with the slightest shrug of their shoulders. Could I do any less? Then I thought of yet another positive. Previously for Yuri and Yuri and now for me, the positive was that our extensions came after we had already passed the halfway mark of our stays. The extensions did not put us in a position where we would once again be in the "first half" of our stay on Mir.

Tumbleweed

Psychologically, it did not seem like as long of an extension as it would have seemed if we had to pass through the halfway point again.

All these ideas were spinning in my head, but the overriding thought was for the crew of STS-79. Of course, there needed to be an extension—there was no way my friends should launch with any unanswered questions about the SRBs—but still, I could feel their pain of having been delayed. I should know.

The first shuttle flight I had been assigned to was STS-51-D, which was to retrieve the Long-Duration Exposure Facility (LDEF). Immediately following our three-week prelaunch press conference, we, the crew, went back to our offices and found out that the shuttle flight manifest had been juggled so that we were not launching in March as the crew of STS-51-D but in June as the crew of STS-51-G. That was only the first of many space shuttle delays I would experience. There was one delay because, the night before launch, the payload center in California couldn't provide support due to an earthquake. There were weather delays, and a delay once occurred because the launch sequence was stopped at T-minus thirty seconds. So, I knew about delays. I could feel the disappointment of the STS-79 crew.

But the bottom line was that I was glad the decision had been made to delay the launch of STS-79. There was no way I wanted my friends—Bill Readdy, Terry Wilcutt, Jay Apt, Tom Akers, Carl Walz, and John Blaha—to launch to come and get me when there was any question whatsoever about the functioning of the SRBs. We were not just talking shuttle crew here; we were talking good friends.

After NASA officially made the decision to delay my return, Frank Culbertson, the head of the Phase 1 office, negotiated with the Russians for comm time so that he could personally tell me the news. I knew that Yuri and Yuri were watching me while I was talking to Frank. The three of us had already talked extensively about Shani's email messages. But my conversation with Frank was different. This was an "official" announcement. I knew they were wondering how I would react. When Frank's voice faded into static, I took off my headphones, hung them up, and said, "Ten years from today, it will make absolutely no difference that I spent an extra two months on Mir." They agreed with me, and we continued our work for the day. I did not, however, resume my practice of counting bags of sou

CREW EXCHANGE

Yuri and Yuri had been on Mir for almost six months. I had been there for five. The Soyuz that had brought Yuri and Yuri to Mir had also been docked to Mir for almost six months. As I've noted, the in-orbit lifetime for the Soyuz was six months. It was time for Yuri, Yuri, and their Soyuz to head earthward. But before they could depart, a new Russian crew had to arrive. One of the checks in the "plus" column when my stay on Mir had been extended was the fact that because of the extension, I would now be able to participate in a crew exchange.

Our discussions as we floated around the table in the Base Block at our mealtimes now centered on the new crew. We endlessly speculated on why the two Russians of the original crew had been replaced by their backups. Yuri and Yuri prepositioned all the bright-blue CNES bags that had arrived on the last Progress at convenient places around Mir so that Claudie, the visiting French cosmonaut, would be able to immediately begin to work when she arrived with the new crew.

One night at supper, Yuri Usachov announced that the ACУ (the Russian acronym for "toilet," pronounced "ah-sue") would not be functional for the next few days.

"What!" I exclaimed in mock horror, hoping that I had not understood his Russian correctly. "Why on earth not?"

"The ground wants me to change out all the parts so that everything will be really nice for Claudie when she arrives."

Tumbleweed

"OK, that is good, but let's talk a little about this. Let's plan the parts exchange in stages so that we can have an ACY for a few hours!"

We continued our discussions and worked out a plan so that all the part exchanges could be accomplished, and we could still, periodically, have access to the toilet.

After all work on the ACY was complete, I decided that maybe a new and longer strip of Velcro would be appropriate on the folding door to the toilet area. Such a strip would allow the door to be more securely closed and ensure privacy when Mir doubled its number of inhabitants. I completed my renovations, and as I floated back to the Base Block, I picked up a big hammer that had just drifted by me in the transfer tunnel. My intention was to return it to its proper place in the toolbox, but as I floated up to the table, Yuri Onufriyenko asked worriedly, "Shannon, why do you have that big hammer? What have you been doing?"

I laughed and, seeing a golden opportunity for a little fun, answered, "Well, Yuri, I have just completed the most important repair we have ever done here on Mir!"

"Where? What did you do?"

I just shrugged and nonchalantly said, "It is finished, and I did a great job! You will just have to find it if you want to know!"

With a look on his face that bordered on panic, he leaped through the air and shot out of the Base Block and into the other sections, hunting for my repair. Yuri Usachov turned to me and said, "Tell me, Shannon, what did you do?" I laughed, repeating my statements about having done the most important Mir repair yet.

I felt a slight twinge of remorse at their extreme reaction to my statement. After all, I had noticed that for the last few weeks, Yuri and Yuri had been showing all the signs of suffering from extreme get-home-itis. One symptom was that they were hypersensitive to anything that might upset our precarious balance here in our cosmic outpost. Things had gone very well our entire time together on Mir. The time until their departure was short. They wanted nothing to happen that would mar the great track record of the Skiffer crew. I sensed this, and I had intentionally played off their paranoia with my statement about the "important repair." When Yuri Onufriyenko returned after searching all the nooks and crannies of Mir without finding evidence of my repair, I promptly, and with much laughter, explained my

Lucid

"repair." They both laughed with relief, and we continued to laugh all evening as we retold the incident with various embellishments.

Anticipation

I thought that the anticipation of the arrival of a new crew would generate a lot of additional activity, but it didn't. After the marathon of changing out parts of the waste management system, we did nothing else special until the morning of the docking. When it was only hours away from arrival time, Yuri Onufriyenko said, "Shannon, where do you think would be a good place for Claudie to sleep and to keep her things?"

I replied, "I think Priroda would be the best place because it is nice and cool. I'm not doing much work in there now since the NASA science program is basically complete."

Yuri and Yuri both agreed, then asked me to fix a nice place for Claudie with all her personal things that had arrived on the last Progress so that she could find them and feel welcome. I rummaged through the nooks and crannies of Mir. I found the new sleeping bag that had arrived for her on the last Progress, some no-rinse shampoo, and a few of the NASA drink bags containing the body rinse that we had been using for our personal hygiene. I also found the blue bag that had come up on the same Progress containing her personal clothes and pictures of friends. I then secured my tape recorder and a selection of music tapes by her sleeping bag so that she could listen to music while going to sleep, and last of all, I left two packages of hoarded cookies stuck in the bungee cord by her things. I called Yuri and Yuri in to inspect my handiwork. They smiled and pronounced it good.

Then the discussion turned to what flags we should display while Mir was the home of two crews from three nations. Yuri Usachov pulled out the French flag that had arrived on the last Progress. It was gigantic. It was six times the size of the Russian and American flags we had been using. We obviously needed flags of roughly the same size. The Russian and American flags could not be made larger, but in a moment of inspiration, I told Yuri and Yuri that the French flag could be folded smaller to make it the same size as the other flags. After all, it was just three bands of color, and the folding could be done so that all three bands were still visible. The Yuris agreed, and then we tried to figure out how we could place Velcro to hold the French flag in the smaller shape. Our schemes didn't work, so I suggested that we sew it to the smaller size. Because it was my suggestion, it became my task. I quickly got busy sewing the French flag so that it would be ready for the Frigate crew when they opened the hatch of their Soyuz.

I had just sewn the last stitch when we caught our first glimpse of the approaching Soyuz. It was hard to accept the fact that inside that faint "star" were three human beings. At the moment of visualization, radio contact was established. As the Soyuz approached closer and closer, the new crew talked about how cold they were in the Soyuz and how they were really looking forward to something hot to drink. As we quickly hung our flags, we told them that we were getting everything ready for them. We certainly did not want to peak too early! Then Yuri stuck the headphones up to the boom box so that they could hear a few strains of a Russian folk song before they started the docking approach.

Arrival

The Frigate crew docked to Mir. On the other side of the closed hatch, we, the Skiffer crew, waited and waited. I will admit, I had certain feelings of trepidation because it had been almost six months since I had seen a human being other than Yuri and Yuri. I wondered if I would remember how to interact with a larger group of people! We prepared hot bags of tea and coffee for the new crew as we waited for them to finish completing their after-docking checks.

Finally, the hatch started to swing open. Yuri and Yuri crowded up to it. I hung slightly back. I had interacted with both Valeri and Claudie while living in Star City but had never seen Sasha before. The Frigate crew floated clumsily out of the Soyuz hatch. They looked like inflatable dolls that had not yet been blown up. My first thought was, "How clumsy they are!" Until that moment, I had not realized how the Skiffer crew had adapted to microgravity life and how graceful we looked as we skimmed around Mir.

I was right alongside Yuri Usachov as he enfolded Claudie in a huge bear hug and exalted, "Oh, Shannon, how glad we are to see you here! Welcome!"

Seconds later, Claudie was embraced by Yuri Onufriyenko, who had not heard the other Yuri's greeting because of the level of background noise and gave an identical greeting: "Oh, Shannon, how glad we are to have you here on Mir!"

Lucid

Sputtering with laughter, I told Yuri and Yuri, "All women are not named Shannon!"

Slightly taken aback at what they had done, they said that it was a good thing they'd had the opportunity to welcome Claudie and make such greeting mistakes with her before they got back to Earth and hugged their wives for the first time!

After we finished greeting each other, the first order of business was to show off our home. Sasha needed no introduction to Mir. He had previously spent five months living and working on Mir, and he was now exploring with interest the new additions, Spectra and Priroda. Valeri and Claudie were experimenting with floating movements in the larger volume of Mir as compared to the cramped space in the Soyuz.

We, the Skiffer crew, the proud hosts, then invited the Frigate crew to the supper we had prepared for them. It was fun to watch them eat, with gusto, the same food that we had eaten for the last six months. Yuri Usachov had found a prized package of American macaroni and cheese and proudly offered it to Claudie. Nothing but the best for our visitors! All six of us tried to float around the Base Block table and eat together, but it was extremely crowded, and invariably, someone was jostled to the background.

The first task on Mir, even before eating supper, was to exchange my Soyuz seat liner with Claudie's. With the arrival of the Frigate crew, I was now officially a member of that crew, and Claudie was officially a member of the Skiffer crew, at least for entry purposes. We quickly exchanged the seats. From that point on, if there had been an emergency that would have forced us to evacuate Mir, I would have flown to the Frigate Soyuz and returned with Valeri and Sasha, whereas Claudie would have gone to the Skiffer Soyuz and returned with Yuri and Yuri.

As we were exchanging the seats, I asked Yuri and Yuri if they had ever done any training, such as entry simulations in the Soyuz simulator, with Claudie back in Star City. They said no. I thought this was a little strange. Why would Claudie do all her landing simulations with a crew she had no plan to return to Earth with and none with the planned team with which she would reenter? I did all the landing simulations with Yuri and Yuri, with whom I would have only returned in an emergency. When I asked Yuri and Yuri about the logic of this,

Tumbleweed

they just shrugged their Russian shrug and said that they had asked the same question back in Star City and had not been given an answer.

As part of the seat-liner transfer, a set of weights that had been individually calibrated for each of us was also transferred. While doing this, I was talking with Claudie, and she told me that during the last few weeks, she had almost not been certified for flight because she did not weigh enough. I said, "How could the doctors have said that? You weigh the same that you always weighed, and you have been there training forever. Why did this issue of weight come up at the last moment?" The answer she had been given was such a typical Russian one. She had been certified to train. She weighed enough for training, but then she needed to be approved for flight, and it was for flight that she did not weigh enough. A few extra weights added to her set fixed the problem, and she was certified to fly.

The first day of the six of us living together on Mir was very interesting, as I rapidly realized that the comfortable rhythms we had lived with for over five months were gone. I laughed to myself when I was thinking about this, and suddenly, the image of a band warming up came into my head. That is exactly how I would describe the first few days of the six of us living on Mir. We were all doing our thing, but we weren't producing a rhythm and a melody together. Things rapidly sorted out, though, and a new rhythm was established. At first, Yuri and Yuri and I had tried to have all six of the Mir inhabitants eat our meals together. This just did not happen. There were several reasons. One was that we just could not all float comfortably around the Base Block table. The other reason was that the Frigate crew was consumed with helping Claudie accomplish the French mission. She was to be on Mir for only fourteen days, so the pace of their work and life resembled a shuttle Spacelab mission and not a long-duration mission. Even though I was officially a member of the Frigate crew and not the Skiffer crew, Yuri, Yuri, and I just sort of slid into our established patterns of eating and socializing together.

Thinking about the arrival of the new crew and the crew exchange, I had anticipated the crowding. I had also figured that there would be certain feelings of dislocation as the group of six realigned into new crew groupings, but the emotion that caught me by surprise was the feeling of loneliness. For the first time since I had left my home on planet Earth, I felt lonely. Thinking about this and trying to figure out why I should feel lonely now, when Mir was more crowded than ever,

I realized that for the first time since I arrived on the space station, I really had no function. Claudie was busy every second working on the French science program. When Valeri and Sasha were not working with Yuri and Yuri, they were working with Claudie. I had finished all the NASA experiments and had also completed packing all the data and equipment that were going to return with me to Earth on the shuttle. Many of my comm passes with the NASA support people in mission control were now given to the French. Any work that needed to be done on the Mir systems was now being done by Valeri and Sasha so that Yuri and Yuri could familiarize them with all the quirks and peculiarities of how we were getting everything to run. I was just the hanger-on, the extra. For the first time on Mir, the days crawled by. I found myself looking forward to the time when I would be living once again, on Mir, as a crew of three.

I was also having a difficult time exercising because most of Claudie's work was being done in the Base Block right by the treadmill and the table. Whenever the space was relatively free, Yuri and Yuri had first priority on the exercise equipment because their return to Earth was imminent, and they needed to make sure they were ready.

The extra folks on Mir really increased the heat load. One day as I was starting to run on the treadmill, I used a temperature strip from the NASA medical kit to determine the air temperature. It was ninety-eight degrees Fahrenheit, and the humidity felt like a Houston summer day. I thought to myself, "This is crazy! How can Yuri and Yuri get ready to go home in conditions like this?" I also wondered how the experiments that Claudie was carrying out could be definitive regarding spaceflight and changes due to microgravity because any changes that would be seen in cardiac function due to microgravity would be overwhelmed by the changes being brought about by the extreme humidity and temperature of the environment.

Saying Goodbye

Suddenly, it was the evening before Yuri and Yuri were heading home. Valeri, Sasha, and Claudie were in Priroda recording a public relations film for the French space agency. Yuri Usachov found me and asked me to come into the Base Block for a Skiffer crew "last tea break." We gathered around the table in the Base Block and reminisced about our time together. In anticipation of heading home, Yuri and Yuri had cut each other's hair. I was joking with them about how short it was and

Tumbleweed

then realized that it was shorter than either had anticipated it being. I stopped the "bad hair day" jokes. We talked about how excited their families would be to see them tomorrow and how much their children would have changed over the course of the six months they had been gone.

Then they presented me with an EVA glove cover that they had used while they were doing their EVAs. They had written on it with a black Magic Marker, "Thanks for your help," and signed both of their names. I protested that I could not accept such a gift. I knew that these glove covers were highly prized by the cosmonauts because they could roll them up small, squeeze them into their returning Soyuz, and once back on Earth, sell them for a good price to collectors of space artifacts. They insisted that I keep it. They said that I had been an integral part of the crew and that the EVAs would not have been as successful without my help.

After accepting the glove cover and thanking them, I rushed into Spectra and retrieved the cardboard "medals" that I had made for each of them. Why cardboard medals? During the Summer Olympics, we received a broadcast on Mir of some of the highlights. The following Sunday evening, after I had done my weekly cleaning of the can opener, we—that is, Yuri Usachov and I—decided that we needed to have a bit of "cosmic Olympics." We both hit on the ideal "game" simultaneously. We would start at the Soyuz hatch, which was docked at the far end of the Kvant 1 module; kick ourselves off; and then furiously propel ourselves through the Base Block, the transfer tunnel, and on to the airlock hatch in Kvant 2. After tagging the hatch, we would flip over and race back to the Base Block table. We would time each other with a stopwatch. It would be Russians versus Americans, male versus female.

We leaped into the competition with gusto. No matter how hard I tried, Yuri was always a few seconds faster. Yuri Onufriyenko arrived in the Base Block just as I was careening to a halt, legs akimbo, amid a shower of floating cameras and tools kicked loose from their Velcroed positions on the walls by my ricocheting trajectory through the Base Block. "What are you doing?" he demanded. We explained the contest and laughingly invited him to participate. I, the lone American, was willing to take on two Russians. "No," he said, looking at the floating cameras and tools and then at my elbows, skinned and bloodied by banging into all the extra equipment deployed throughout Mir. "This

Lucid

is not a suitable activity for us to be doing! Why, you might kick through the outer hull and cause a leak!"

We stopped, Russia still holding the lead. I thought to myself, "Sometimes it is just not much fun being the commander." I thought back to one of my shuttle flights, during which we played "space soccer" with a rolled-up pair of socks. The object was for everyone on the middeck to kick the "sock ball" into the airlock while holding their arms behind their backs. Yes, it was fair game to kick the other crew members away from the sock. You could use your head and your feet, just not your arms. As soon as the going got fast and furious, with crew members literally bouncing off the walls and the ceiling, the commander floated down from the flight deck, took one look, and called the game to a halt. As I said, sometimes it is not much fun to be the commander.

Anyway, after these games, we discussed what other cosmic Olympic events we might do and the design of the medals that we should receive. So, when it was approaching time for Yuri and Yuri to leave, I naturally wanted to send something home with them to express my appreciation of our time together, and the concept of a space medal popped into my head. I removed the lightweight cardboard cover from one of the NASA procedure books, cut out medallion-size stars, covered them with the metallic tape we used for repairs, made short red and blue streamers from red and blue tape, and then wrote "#1 Commander of the Universe" on one and "#1 Board Engineer of the Universe" on the other. As soon as I presented Yuri and Yuri with their medals, the others floated into the Base Block, and it was back to accomplishing mundane tasks like collecting the trash left over from supper.

It was departure morning. Everything had been done. We took the final pictures of all of us in various crew combinations: the two women of Mir, the new Skiffer crew, the new Frigate crew, the old Skiffer crew. We hugged each other, and then Yuri, Yuri, and Claudie floated into the Soyuz and slid the hatch shut. This left Valeri, Sasha, and me alone inside Mir. For a few minutes, Valeri, on the Mir side of the hatch with his headset on, did the final leak checks with Yuri and Yuri. Then that was finished, and to us, Yuri, Yuri, and Claudie were gone from our world, even though they had not undocked. On Mir, the Frigate crew was starting a new rhythm of daily life together. The Soyuz undocked.

A few hours later, we heard them on the radio. They told us that they had not been able to talk to the Russian control center but that things were normal, their parachutes had deployed, and they should be on the ground shortly. As we floated around the table for lunch, it was odd to think that just a few short hours before, we six had been having breakfast together, and now Yuri, Yuri, and Claudie were enfolded in the arms of their families and enduring the medical checkups and tests of the Russian doctors.

The Frigate Crew
Now I was the most experienced crew member aboard Mir. I found that I liked it! When I had arrived on Mir, Yuri and Yuri seemed to know everything about living there. The thought had even crossed my mind that maybe there had been a separate training program for the cosmonauts as opposed to the American astronauts, because I certainly did not know what they did! Now it dawned on me that there was no separate training program; there was just the experience of living on Mir! Now I was the expert who knew that the long white underwear was stored under the hatch in Kristall and that the extra-large trash bags were under the soft panel in the hatch between the Base Block and Kvant 1. Knowing locations was definitely a power trip! I found it exhilarating to be asked where to find this, how to do that.

After I warned Valeri about the dangers of changing clothes every day because no more would arrive until the next Progress did, he asked me to let him know just how much of every consumable there was on board. I informed him that only two dry towels were left—and that I was saving them for John as an extra-special present when he arrived. There were just enough wet gauze squares to have one a day for bathroom use, but there was enough food to last for a year! The Skiffer crew had barely started eating the food that was sent up for us for our time on Mir. We had been eating the food that was there from before Norm's flight.

Life, for me, now settled into exercising, finishing odds and ends of packing, and helping Valeri and Sasha maintain Mir. One big job that I routinely did was to try various ways to dispose of the water that had started collecting in large pools all around Mir. For several weeks, the pooling had worsened. Now it seemed that we could never open a panel without dislodging a large glob of water that showered all over

Lucid

us or place our hands anywhere to pull ourselves through the corridors without plopping it into another glob of water. The refrain "water, water everywhere, but not a drop to drink" kept racing through my mind. We tried absorbing the water with dirty clothes and towels and then carefully placing the soaking-wet items into bags. We tried sucking it up with a makeshift wand attached to a small vacuum pump. I found using a very large syringe from the medical kit a good, but slow, way to suck up hard-to-reach balls of water. I asked Valeri for a reason for the water accumulation, and I thought he said something about orientation and humidity, but the part I understood was that things would improve, and so they did.

I found everything that needed to be taken to Earth on the shuttle except for the quail eggs that had been fixed so many months before. They were in plastic bags stored in bright-blue metal suitcases, about twelve by twelve by six inches—nice-size boxes. I was missing three. I hunted everywhere. They were not where I had put them and so faithfully recorded the location. This really surprised me because we, the Skiffer crew, had made a deal that no one would move something that had been put in a certain place by someone else without letting that person know and then writing it down. This was especially necessary during the last part of our time together because it seemed that each of our short-term memories had disappeared. It was difficult to remember something without writing it down. In fact, it was so noticeable that I even inquired of Gaylen if there were any heavy-metal contaminants in our drinking water that could be producing these kinds of symptoms. He assured me that there were not and guessed that the lack of outside stimuli was causing the problems with short-term memory. He must have been correct, because the problem disappeared in the crew transport van right after landing.

Not being able to find the eggs really bothered me. I knew that I could not head earthward without them. I looked behind every panel in Specter and Priroda. No blue boxes. I wrote what was behind each panel on a sticky note and stuck these notes on the panels so I would know that I had searched behind them. This was turning into the most stressful Easter egg hunt of my life. In desperation, I decided to open the ceiling panels in Spectra. And there, after shaking the dislodged water from my eyes, I saw the blue boxes wedged in between bags of German science equipment left on Mir from a long-ago flight. I breath

Tumbleweed

huge sigh of relief. Now I was ready to head home, if the hurricanes haunting the Florida coast would let me.

STS-79 LAUNCH, MIR DOCKING AND LANDING

Bill and Gaylen kept me appraised as to the shuttle preparations for the launch of STS-79. After all, we all had a vested interest in it. Soon after my arrival on Mir, Gaylen had told me that his wife was pregnant. Before the delay of STS-79, there was no contest between the arrival of their baby into life on Earth or my return to life on Earth. I would be back long before the baby was born. But the delay of STS-79 changed all that. Our arrival times were now almost the same. I told Gaylen that I could get along just fine with the other crew surgeons who were in Moscow training for the next Mir increments and that he needed to be sure he was back in Texas well in advance of his baby's birth. I reminded him that the wonderful comm techs in Houston could patch him in to me from the phone in his home for our weekly private medical conferences. He agreed and returned to Houston at the end of August. Then, it was just Bill and me holding our breath in anticipation of returning back home to our families in Houston.

After *Atlantis* was brought out to the pad in preparation for launch, I would try to get to the window in Kvant 2 whenever we made a daylight Florida pass. With the high-powered Russian binoculars, I was able to see the shuttle—three gleaming white pencils, two skinny, one fat—standing on end on the launchpad. I could also see a huge hurricane in the Atlantic. It came as no surprise when Bill informed me that the shuttle was going to be rolled back into the Vehicle

Tumbleweed

Assembly Building (VAB) and that the STS-79 launch had been delayed again.

After a couple of days, it was back out on the pad, and the countdown began again. A few days later, I looked down at the Atlantic with utter disbelief as I saw another hurricane headed to the cape! Unbelievable! Once again, the shuttle was rolled back into the VAB. Never before had the stack been rolled back twice in one shuttle-launch flow due to hurricanes!

I must admit that I made a conscious effort not to let my imagination roam over all the possibilities of what could go wrong with my ride home as it was being hauled back and forth between the VAB and the launchpad. One night, I had a nightmare about the shuttle toppling off the vehicle that transports the shuttle stack to and from the pad. I also began to wonder just what the plan would be if the shuttle did not arrive. Would I return in the Soyuz? How long would I remain on Mir before the OK would be given for me to return in the Soyuz? Valeri, Sasha, and I discussed different scenarios and the ramifications of each delay endlessly. "Shannon getting home" always made for interesting supper conversation!

I kept getting pithy emails from Michael. He would always include a statement such as, "There are only seventeen Russians who have been in space longer than you have. It is time for you to come home."

Or, "If you stay in space just 365 more days, you can beat the record of the longest-staying Russian!"

Or, "Remember the chimps in space who went crazy? We don't want you home if you go crazy like they did!"

In one email, I asked my daughters if they had bought new outfits to wear at the landing. (It was a family tradition for me to buy them each a new outfit whenever I came home on the shuttle.) Their exasperated answer was, "Of course not. We can't pick out something for landing when we don't even know what season of the year you will be coming home in!"

Finally, the day came when there were no more hurricanes out in the Atlantic, there were no more system problems with the shuttle, and all was "go for launch." The three of us aboard Mir knew the time that *Atlantis* should launch and had endless discussions over whether we would be able to see it. As the time for the launch of *Atlantis* neared, we began surging back and forth to various windows to see if we could catch a glimpse of the liftoff. It was a night pass. It was dark. I was at

Lucid

the window in Kvant 2 when I suddenly saw a new star. It got brighter and brighter. It couldn't be, but I was sure it was. I raced into the Base Block and yelled at the guys to come and see. Sasha's finger flew to his lips, and he said, "Shhhhh . . . listen! What are they saying?"

I stopped midbreath and listened to the crackle of the ham radio. It was mission control and the voice of Bill Readdy, the commander. Some thoughtful ham operator was patching mission control through to us. Intermixed with the crackle, I thought I caught words about an APU (auxiliary power unit) problem and then something about return to launch. I stopped breathing, and then I realized that there was radio traffic about APU 2 shutting down early, and it was the negative return call I had heard, not a call for an RTLS (return to landing site). Valeri and Sasha were yelling at me to translate for them. Was the shuttle OK? I assured them that it was not only OK but headed our way. But I must admit that on that night pass when we were over the Florida Panhandle, I started to let my breath out slowly, slowly, ever so slowly and did not seem to breathe again for days—until I heard and saw the hatch of *Atlantis* crack open. Later that evening, we had a comm pass with the Russian control center, and they confirmed what we already knew—the shuttle was headed our way!

Finally, it was time for us to prepare for the anticipated invasion of the shuttle crew! It seemed to me that the biggest hindrance to my becoming part of the STS-79 crew was the fact that the docking module on Mir was stacked wall to wall with food containers. Unless we could find someplace else to put them, there was no way the Mir hatch could be opened to allow the shuttle crew to enter! It was beyond my imagination where we could store all these containers. Fortunately, Valeri and Sasha did not have my limitations in imagination. They decided to put all the food containers from the docking module behind the panels in Priroda. For two days, we spent all our spare time ferrying food containers into Priroda. We formed a human chain and passed the containers into the module, where we stuffed them behind the panels. The only disadvantage to our solution was that from then on, I had to remove all the interfering food containers before I could relocate any NASA experiments from one panel to another. After we completed the switch, there was nothing to do but wait.

Hours before the planned docking time of STS-79, we began looking out the window to see who would see it first. Then we saw it far beneath us, a large white dissected airplane, slit down the spine, gutted and cleaned. It ever so gracefully approached us from below, suspended over the blue-and-white background of Earth.

Mir was the passive partner in this celestial ballet, so we, the Mir crew, had no active role to play. Our job was only to watch and wait. Finally, the *Atlantis* docked. On the Mir side, we felt nothing; we only knew docking was complete because of the radio calls confirming it.

Valeri suddenly fell into host mode and decided that we needed to get ready to greet the arriving crew. He told us to find some bread and salt. He assured us that the Americans would be expecting this traditional Russian welcome. Quickly, we found a round metal lid from some long-ago experiment that vaguely resembled a plate and then rifled through food containers to find six bread packets. But what should we do for salt? I suggested the salt tablets from one of the many green Russian medical packets on board. We taped the salt tablets and bread packets to the "plate." Then Valeri had an inspiration. He retrieved the dried reddish-yellow autumn leaf that he had found the previous evening as he was skimming through one of the many Russian books on Mir. At the time, we had commented that someone must have sent it up on a long-ago Progress to remind a former crew of the changing seasons back on Earth. This leaf was also taped to the plate. We were now ready for visitors!

After what seemed like an endless wait, the hatch separating *Atlantis* and Mir was opened. The STS-79 crew surged into the Mir docking module. Valeri and Bill, the two commanders, bear-hugged, slapping each other on the back. I hung back just a trifle. True, these were my friends, but all of a sudden, faced with the reality of in-person contact with them, I instantly felt shy. I even had the fleeting thought, "Will I be able to speak and understand their language?" I thought about not having had a shower for six months. Did I smell? I had on the same blue jumper that I had worn every day for 188 days and the Russian boot liners that I loved to wear. I felt like the country cousin mouse when the city cousin mouse unexpectedly pops into her humble country home. "Don't be so silly," I admonished myself, then enthusiastically joined in the generalized hugging.

It was all babble and confusion. Somehow, we managed to get back into the Base Block for the standard "we are so glad to be here" presentation to mission control. As soon as that was finished, I floated into the shuttle. It seemed so white, so sterile, and yes, so American. I had experienced the same overwhelming feeling of brightness after disembarking at the Atlanta, Georgia, airport on my first return trip to the States after living in Russia. The absence of drab was overwhelming! It certainly did not have the homey, lived-in look that I had grown accustomed to in Mir. I quickly floated up to the flight deck. Oh, those shuttle windows! So large! The view of Earth was just so overwhelmingly awesome after the smaller windows on Mir. "I could just float here forever," I thought.

But then, no utterly sublime moment lasts forever, and my mind immediately recalled the prosaic reason that had brought me so quickly into the shuttle. I was ready for clean American clothes! I was ready for some shuttle shorts and Lands' End shirts! I rushed to my clothes drawer. Oh yes, there was underwear. Underwear! What a novel concept! A clean pair for each of the days that I would be on the shuttle! This seemed excessively extravagant to someone who had gone six months without! There were clean shorts, one for each of the shuttle days, but where were the shirts? I pulled everything out of my drawer. No shirts. I went to each of the STS-79 crew members and asked if, by chance, they had any extra shirts in their drawers. The answer was no. I finally had to face up to the fact that no one in the bureaucratic NASA system had remembered to pack me any shirts. This just proved once again the validity of the Lucid family rule: do your own packing! When the shuttle crew realized that I had no shirts, they laughingly each gave me one of theirs.

I helped Sasha bring the large bundle that contained John's individual molded seat for the Soyuz over to Mir. It was quickly exchanged for mine. John became an official member of the Mir crew, and I was officially a member of STS-79 and was headed home! First, though, there were days of timeline tasks to work through. I attached myself to Tom Akers and told him that I was his unofficial helper. I was his extra pair of hands. Whatever he would tell me to do, I would do!

The next few days blurred together. There were several thousand pounds of supplies to bring over to Mir and all twenty-one of my packed bags to bring to the *Atlantis*. I ate and slept with the shuttle crew in the shuttle. But I was still really partial to the bathroom facilities in Mir. I asked Valeri if I could continue to use the restroom in Mir as long as the shuttle was docked to it. With a good-humored laugh, he assured me that I would always be welcome in the Mir bathroom. In the evenings, after all the work was done, the Mir and

Tumbleweed

shuttle crews would gather around the table in the Base Block. I had to laugh because now a lot of the conversation centered around sports events. I told the STS-79 crew that there was a lot to be said for being part of a Russian crew. For the six months that I had been part of one, I had never once been subjected to a football, baseball, or basketball conversation.

The appointed time for us to undock finally arrived. We hugged the Mir crew goodbye in the docking adapter, but John was in Priroda, working on an experiment. I told Bill Readdy not to close the hatch until I got back and flew in there to say goodbye. Then it was back to the hatch, and it slid closed with a sigh. Although still physically joined, we were now two separate worlds. The *Atlantis* crew was focused on returning to Earth, and I knew that the Mir crew was focused on what needed to be done that day. For them, tomorrow would take care of itself.

The shuttle crew went through the undocking procedure, and we slowly backed away from Mir. Gazing out of the big shuttle flight deck windows, I had mixed emotions as I waved goodbye. For six months, Mir had been my home. I knew every nook and cranny. Three friends were on board. I had a small sense that I was deserting them by leaving now. I thought of Sasha being there for five more months without any Jell-O. While I was on Mir, right after Yuri and Yuri had left, Sasha and Valeri asked me about Jell-O. They had heard that we, the Skiffer crew, had it every Sunday night, and they wanted to make sure that the Frigate crew was going to continue that tradition. I assured them that there was enough on board for us to do that.

Just before *Atlantis* docked, we ate the last bag. I assured the guys that there would be more aboard *Atlantis* and that I would make Sasha the Jell-O czar. I would find it on the shuttle and then make sure that Sasha knew where it was stored on Mir. One of the first things that I did after *Atlantis* docked and the hatch opened was to find the food containers aboard the shuttle. I rummaged through them, searching for the one that would contain the Jell-O. I searched and searched and then, with disbelief and sorrow, had to inform Sasha that there would be no Jell-O for them after I left.

Despite this small regret, I gazed at Earth and kept thinking of what it was going be like to step foot on its surface again and feel the wind in my face and the embraces of my family. I touched my pocket where

Lucid

the last poem that Kawai had sent to me was safely kept and repeated
the memorized lines softly to myself:

When I Get Home from Space

When finally, I get home from space,
There will be many things to do;
I know that it won't take too long
to get back in the groove.

I'll hug my family and greet my friends,
then sit down in my chair.
With any luck, a brand-new book
will be waiting for me there.

My family may have bought for me
a cookie of chocolate chip.
We'll gather 'round and have a bite
and sun tea we will sip.

Once I've rested and gotten used
to being normal weight,
(And knowing me, I'm sure it'll be
sooner and not late!)

I think I'll find my rollerblades
and skate over to the mall.
I'll spend ten hours at Book Stop,
and bring home quite a haul.

For breakfast I'll have IHOP,
for lunch and supper too.
or maybe I'll eat at Black Eyed Pea
or try steak at someplace new.

I'll want to take a bike ride,
my family might come along.
We'll ride to Armand Bayou,
and nothing will go wrong.

Tumbleweed

Later, I might bake a cake
in the new oven my husband bought.
I may even cook meals for my kids
to tell them "thanks a lot!"

And in my study on Sunday noon,
I'll be reading in the room,
then take a lengthy midday nap,
and wake, rested, for more fun.

Deorbit morning arrived. On previous shuttle flights, my responsibility had been to get the middeck ready for reentry and to help the flight deck crew suit up in their launch-and-entry suits. I am not boasting when I say that this was a particular talent of mine. So on deorbit morning, I asked Carl Walz, the STS-79 crew person in charge of the middeck for deorbit, how he had it orchestrated in his mind, and then I just jumped in to help him. We worked together as a team. You would have thought we had practiced this scenario many times together rather than this being the first time we had ever done it jointly. We were a great team. Little did we realize as we worked together that in five years, the crew Carl would be part of, Expedition 4 on the ISS, would be in space for 190 days. This was longer than the 188-day record I had just set for the longest duration of an American in space. Thus, he would become one of the new record holders for a period of time.

During this time of deorbit preparations, shuttle crews always fluid-load—that is, they drink lots of water to replace that which the body had sensed as extra fluid and removed while living in microgravity. Carl and I filled drink bags for everyone and placed them where they could be reached. Then we put up my seat. It had been decided that long-duration crew members would return in a recumbent position—that is, lying on their backs. This was to prevent the crew member from fainting while the body readapted to working against gravity pulling blood down into the lower part of the body. Just like every time I had returned to Earth, I prepared my "nest" by placing the drink bags around me where they could be easily reached. I also taped salt tablets within reach. I put a few packets of cookies in strategic places so that I could eat them and make myself a little thirsty to drink more fluid.

Lucid

I was a little late strapping into my seat and felt myself scrambling as I configured my nest. I laughed as I looked at myself—surely this was an inglorious way to arrive back on my planet! I was strapped into my seat, lying on my back, with my boots stuck into an empty middeck locker space. Down on the middeck, we had nothing to do except wait for the flight deck crew to get us home.

Just after Bill rolled the *Atlantis* to a stop on the shuttle landing strip at Kennedy Space Center, the Houston capcom welcomed us home, adding a special welcome to me. I tried to answer, but my comm box had slid onto the floor, and the moment had passed by the time I retrieved it. I sat up and tried to get my helmet off. It was stuck fast and would not budge. Then the hatch opened, and suit techs came in. I explained my problem with the helmet, and they quickly got a screwdriver and went to work. Finally, they were able to pop it off. As I was standing up, the two flight docs came in and asked how all of us were doing. One went to determine how the rest of the crew was doing, and Gaylen, who had flown from Houston to KSC to be my flight doc after landing, stayed with me. We were just standing there, and he kept asking me how I was doing, and I said I was fine. He checked my blood pressure. It was normal. He mentioned something about being sure not to release the *g*-suit while I was sitting down. I remembered that I had forgotten to pin the *g*-suit, so it had never inflated. Oh well, I was doing fine without it!

Finally, I suggested we get moving. The suit techs stuck their heads in and asked if I wanted to be carried off. "No, thanks, I am doing fine," I replied. I lurched out of the middeck and into the crew transport van. Several thoughts were running through my head concurrently. One was that my locomotion seemed as good as it had been every time I had returned from space, and the other was the sense of being overwhelmed by the number of people and the "vastness" of the crew transport van.

After what seemed like an eon, the transport vehicle finally arrived at crew quarters. And then, at last, I was able to see my family. They were all there: Mike, Kawai, Shani and Jeff, and Michael. And no, there were no great philosophical exchanges at that moment, our first time together on the same planet in six months. Michael, always eloquent, summed up the magnitude of the moment for us all. "Wow, your arms look so skinny, Mom, and your hair is as gray as Rich Clifford's!"

It was *so* good to be home!